E FOR

**Site Investigations in
Areas of Mining Subsidence**

Site Investigations in
Areas of Mining Subsidence

Edited by

F. G. BELL, B.Sc., M.Sc., Ph.D., C. Eng., M.I.M.M., M.I.Min.E., F.G.S.

LONDON

NEWNES – BUTTERWORTHS

THE BUTTERWORTH GROUP

ENGLAND
Butterworth & Co (Publishers) Ltd
London: 88 Kingsway, WC2B 6AB

AUSTRALIA
Butterworths Pty Ltd
Sydney: 586 Pacific Highway, NSW 2067
Melbourne: 343 Little Collins Street, 3000
Brisbane: 240 Queen Street, 4000

CANADA
Butterworth & Co (Canada) Ltd
Scarborough: 2265 Midland Avenue, Ontario M1P 4S1

NEW ZEALAND
Butterworths of New Zealand Ltd
Wellington: 26–28 Waring Taylor Street, 1

SOUTH AFRICA
Butterworth & Co (South Africa) (Pty) Ltd
Durban: 152–154 Gale Street

1001032627

First published in 1975 by
Newnes-Butterworths, an imprint
of the Butterworth Group

ISBN 0 408 00164 X

Text set in 11 pt. Photon Times, printed by photolithography,
and bound in Great Britain at The Pitman Press, Bath

Preface

With the exception of London and Merseyside, most of the large conurbations in Britain are located on coalfields, in some of which mining began centuries ago. Usually few, if any, records have been left of the earlier mining operations, which of course, were at shallow depth. Now that urban areas are being developed and redeveloped, old mine workings often pose one of the worst problems an engineer has to contend with. Furthermore, mining is still proceeding in many urban areas, but the problems which modern longwall mining give rise to as far as ground engineering is concerned are more easy to tackle because of the predictability. If old mine workings are suspected at a site which is to be developed then a thorough site investigation is called for.

This text outlines how site investigations may be carried out as well as reviewing mining methods. It also makes some suggestions concerning the methods by which foundation problems caused by mining operations, either past or present, may be dealt with. Accordingly, the text should prove of use to all engineers who are working in mining areas.

F. G. Bell

Contents

Chapter 1

Introduction

Site Investigation

The need for urban redevelopment on a large scale together with the increase in the rate of construction due to increasing mechanisation and the increasing scarcity of suitable sites has meant that in recent years areas formerly regarded as unsuitable have been, and are being, considered for building purposes. As a result, ground treatment techniques which in the past were looked upon as remedies of last resort for dealing with unforeseen problems connected with unsuitable ground, are now being incorporated into construction programmes. Moreover, because construction programmes are now more rigid, delays in below-ground work can prove even more costly. All this means that the need for thorough site investigation is vital if ground problems are to be recognised and the correct treatment diagnosed.

Most of the large industrial centres of the UK, in all of which redevelopment is going on, are underlain by rocks of Coal Measures age. An added factor as far as site investigations in such areas are concerned is the problem of past or existing mineral workings. It must not, however, be assumed that the frequent problems associated with mining in these areas are only related to the extraction of coal, for other materials have been and are won from the Coal Measures. These include fireclay; gannister; ironstone; clays, shales and mudstones for bricks; sandstones for building purposes, etc. Such materials have been both mined and quarried.

Ground covered by discard from old mines is now being developed with increasing frequency. Mine waste may be poorly compacted and in such cases has a low strength. If heavy structures are founded upon it they are likely to cause appreciable settlement. Site investigations in such areas are therefore necessary to assess the engineering performance of this material.

The object of a site investigation is primarily to assess the suitability of a site for construction purposes and to provide information for the construction design team (Fig. 1.1). It first of all involves a survey of the relevant literature, ordnance- and geological-survey maps and air photographs. In this respect the obvious authorities to consult are the Institute of Geological Sciences, with offices in London, Edinburgh, Leeds and Exeter, and the National Coal Board area offices. Most industrial areas have been surveyed from the air so that air photographs may be available from concerns like Huntings Surveys, Meridian and Fairey Aviation. Air photographs, however, prove of little use in built-up areas. Local authorities, museums, libraries, mining organisations and firms may also provide valuable information. Nonetheless in some instances it may be more economic to obtain the

1

2

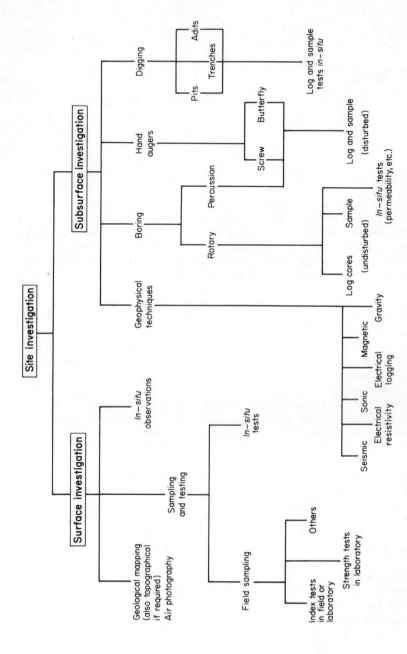

1.1. Organisation of a site investigation (after Fookes, P. G., 'Planning and Stages of Site Investigation', *Engng. Geol.*, **2**, 81–106 (1967))

required information directly from the site exploration rather than proceed with extensive searches through the literature which could yield little.

In areas of productive Coal Measures, if no record of mining activities is unearthed then it should not be assumed that mining has not taken place. It must be borne in mind that mining has gone on in many coalfields in the UK for several centuries, but that the first statutory obligation to keep mine records only dates from 1850 and it was not until 1872 that the production and retention of mine plans became compulsory. In many mining areas houses have been built over old shallow mine workings and their presence should not be regarded as indicative of the absence of workings. A preliminary visit to the site and its immediate environs provides a view of the setting from which it is necessary to plan subsequent exploration.

An investigation of a site for an important structure requires the exploration and sampling of all strata likely to be significantly affected by the structural load. The location of sub-surface voids due to mineral extraction is of prime importance in this context. Large structures involve excavation. As a result not only is the knowledge of the sub-strata necessary, but also that of the ground water conditions is needed. The presence of old shallow mine workings tends to alter sub-surface drainage regimes. The type of sub-strata and ground-water conditions affects the ease with which such material can be removed, and the stability of the sides of the excavation. They are also significant controlling factors as far as earth pressures and the bearing capacity of the foundation rocks are concerned.

Site exploration involves both the exploration at and below the surface. For sub-surface examination at shallow depths, and where conditions prove suitable, trial pits and trenches allow visual inspection as well as mapping and sampling from continuous areas. Such excavations can be quickly made with mechanical diggers, but due regard must be paid to safety precautions. Their location and number will largely be governed by the nature of the site and the type of structure to be erected.

The location and depth of boreholes also depends upon the nature of the ground conditions and the shape, extent and load of the structure. Suffice it to say that holes should be located so as to detect the stratal sequence, the thickness of its individual members and its geological structure. Exploration should extend to a depth which includes all strata which are likely to be significantly affected by the structural load. If the presence of old mine workings is suspected then their detection, type, extent and present condition must be ascertained. Probe boring using rotary percussion or rotary methods will be required to provide the answer (Fig. 1.2). If old workings are present then once their basic pattern is established the details can perhaps be filled in by using hand drills.[1]

Examination of below-surface workings can be carried out by using borehole cameras or closed-circuit television. In some instances it may be possible to enter old workings, however, since this could very well prove dangerous, it should not be undertaken except with the co-operation and assistance of the National Coal Board.

During drilling the maximum amount of information should be recorded. In addition to a log of the stratal sequence and its description, the mechanical properties of the rocks may be noted. The simplest, but yet one of the most important, is the fracture spacing. One method of quantifying the fracture spacing is that of the rock quality designation.[2] The *RQD*, as it is commonly referred to, is the ratio of the length of intact rock in *NX* core sections in excess of 100 mm to the total length drilled, expressed as a percentage. The following four qualities are recognised:

Description	RQD (%)
Excellent rock	Over 90
Good rock	75–90
Fair rock	50–75
Poor rock	Under 50

Incidentally, in fractured or soft rock, a double-tube-core barrel is essential to obtain better core recovery. More simply, a fracture spacing index can refer to the distance apart of the fractures in a given length of core for each rock unit. Care must be taken to determine that the fractures are original, i.e. they have not been produced as a result of drilling.

1.2. Probe boring for detecting old mine workings (courtesy D. A. Richardson)

Sampling is an important part of any investigation. Samples should be taken from each rock type for laboratory testing. The latter should take place at the same time as the site investigation. In this way results can be fed back to the site operators, which should facilitate the development of their programme. Undisturbed sampling is necessary for most tests; it is not a problem as far as hard rocks are concerned but it can be a different matter when soft rocks are involved. Several sampling devices are marketed which attempt to solve this problem. Samples must always be meticulously tabulated, and packed and sealed with care.

An assessment of the indirect tensile strength of a rock on site can be made by using a portable point-load tester (see Fig. 1.3). There is generally a good correlation between the point-load strength and that of unconfined compression. Franklin, Broch and Walton[3] have used the point-load strength and the fracture index as the basis for a rock quality classification. Another simple test, again which usually produces a reasonable correlation with unconfined compressive strength, and which is non-destructive, is the Schmidt hammer test.

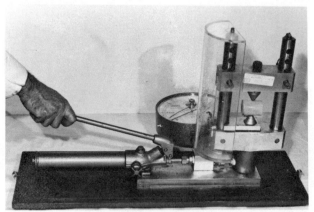

1.3. Portable point-load tester

Rock quality indices should also take account of the degree of rock weathering encountered during a site investigation. This can be based on a simple description of the character of the rock as seen in the field and core samples, the description embodying different grades of weathering which are in turn related to engineering performance (Table 1.1).[4] A number of tests have been designed to assess certain aspects of durability. Perhaps one of the most familiar is freeze–thaw testing. More recently the slake-durability test has been introduced.[5] This test estimates the resistance of argillaceous rocks to wetting and mechanical abrasion and it has been shown that there is a general correlation between slake-durability indices on the one hand, and the rate of weathering and stable slope angles of quarries and pits, on the other.

Swelling of rocks is associated with weathering. Clays, shales, mudstones and marls are most prone to swelling; however, small amounts have been recorded in some sandstones. The clay-mineral content of argillaceous rocks plays an important role as far as swelling is concerned, e.g. kaolinite is not expansive whilst montmorillonite is. When a rock swells it does so due to the absorption of water which allows the development of pore pressures high enough to overcome its inherent strength. Rocks which have an unconfined compressive strength of 41 MN/m^2 and above are not subject to swelling.

Accurate recording of ground-water conditions is important, particularly if excavation level extends beneath the water table. Moreover, the pattern of ground-water flow may be related to old mine workings. Not only should the water levels be observed in boreholes, say twice daily, but at least one standpipe should be installed for long-term observation. Piezometers may be installed in boreholes and *in-situ* permeability tests carried out. In some instances the ground-water may

Table 1.1 ENGINEERING GRADE CLASSIFICATION OF WEATHERED ROCK (after Fookes, Dearman and Franklin[4])

N.I.R.R.*				Field recognition (after Fookes and Horswill, 1969)		Engineering properties (after Little, 1969)
Index† value	Classi- fication	Grade	Degree of decomposition	Soils (i.e. soft rocks)	Rocks (i.e. hard rocks)	
12 / 11	Residual soil	VI	Soil	The original soil is com- pletely changed to one of new structure and composition in harmony with existing ground surface conditions	The rock is discoloured and is completely changed to a soil in which the original fabric of the rock is completely destroyed. There is a large volume change	Unsuitable for important foundations. Unstable on slopes when vegetation cover is destroyed, and may erode easily unless a hard cap is present. Requires selection before use as fill
10		V	Completely weathered	The soil is discoloured and altered with no trace of original structures	The rock is discoloured and is changed to a soil, but the original fabric is mainly pre- served. The properties of the soil depend in part on the nature of the parent rock	Can be excavated by hand or ripping without use of explosives. Unsuitable for foundations of concrete dams or large structures. May be suitable for foundations of earth dams and for fill. Unstable in high cuttings at steep angles. New joint patterns may have formed. Requires erosion protection
9 / 8	Badly weathered	IV	Highly weathered‡	The soil is mainly altered with occasional small lithorelicts of original soil. Little or no trace of original structures	The rock is discoloured; discontinuities may be open and have dis- coloured surfaces and the original fabric of the rock near the dis- continuities is altered; alteration penetrates deeply inwards, but corestones are still present	Similar to grade V. Unlikely to be suitable for foundations of concrete dams. Erratic presence of boulders makes it an unreliable foundation for large structures

			Soil	Rock	
7 6 5	Weathered	III Moderately weathered‡	The soil is composed of large discoloured lithorelicts of original soil separated by altered material. Alteration penetrates inwards from the surfaces of discontinuities	The rock is discoloured; discontinuities may be open and surfaces will have greater discolouration with the alteration penetrating inwards; the intact rock is noticeably weaker, as determined in the field, than the fresh rock	Excavated with difficulty without the use of explosives. Mostly crushes under bulldozer tracks. Suitable for foundations of small concrete structures and rockfill dams. May be suitable for semipervious fill. Stability in cuttings depends on structural features, especially joint attitudes
4	Fresh	II Slightly weathered	The material is composed of angular blocks of fresh soil, which may or may not be discoloured. Some altered material starting to penetrate inwards from discontinuities separating blocks	The rock may be slightly discoloured; particularly adjacent to discontinuities which may be open and have slightly discoloured surfaces; the intact rock is not noticeably weaker than the fresh rock	Requires explosives for excavation. Suitable for concrete-dam foundations. Highly permeable through open joints. Often more permeable than the zones above or below. Questionable as concrete aggregate
3		I Fresh rock	The parent soil shows no discolouration, loss of strength or other effects due to weathering	The parent rock shows no discolouration, loss of strength or any other effects due to weathering	Staining indicates water percolation along joints; individual pieces may be loosened by blasting or stress relief and support may be required in tunnels and shafts

*The N.I.R.R. classification is given for comparative purposes and the grade figures given throughout the text are those in the third column.
†Total index value obtained by adding together the index value for lustre, hardness and consistency and state of crystallisation.
‡The ratio of original soil or rock to altered material should be estimated where possible.

contain substances in great enough quantity to adversely affect cement concrete, the sulphate content and pH value being of particular interest. A chemical analysis of the ground-water is then required to assess the need for special precautions.

Generally speaking, geophysical methods have had only limited success in the detection of old mine workings. The use of the proton magnetometer for shaft location has been described by Raybould and Price,[6] and of resistivity methods to detect fissuring and cavities in limestone by Early and Dyer.[7] The use of geophysical methods is discussed in Chapter 4.

Standard and dynamic penetration tests are of limited value in foundation investigation in Coal Measures areas. Field-loading tests are usually the best method of assessing the strength and deformation characteristics of the rocks concerned. A value of Young's modulus can also be determined from seismic tests. Loading tests can be either carried out on bearing plates or piles, however, because the ground immediately beneath a plate or pile is capable of carrying a heavy load without excessive settlement; this does not necessarily mean that the ground will carry the proposed structural load, especially where a weaker horizon occurs at depth but is still within the influence of the bulb of pressure generated by the structure (Fig. 1.4).

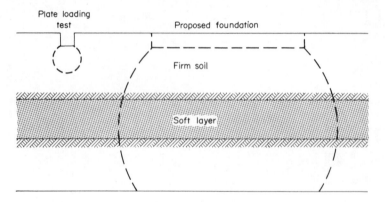

1.4. Bulb of pressure beneath a foundation as compared with one beneath a plate load test

Plate-bearing tests are carried out in trial pits usually at excavation base level (Fig. 1.5*a*). Plates vary in size from 0·15 to 0·61 m in diameter, the size of plate used being determined by the fracture spacing. Usually the maximum loading in a test is approximately twice the design load so that they rarely reach the ultimate bearing capacity at which settlement continues without increasing load. As a consequence it is generally only necessary to show that a foundation is capable of carrying a given load rather than determining the point at which it will fail. The rock beneath the test plate should be examined to a depth of 1 m after the test.

Pile loading tests are usually carried to failure, this giving the ultimate bearing capacity of the particular pile that has been tested (Fig. 1.5*b*). However, to infer that the ultimate bearing capacities of other piles on the same site are similar from such loading tests alone requires either the prior testing of an adequate number of piles and the use of a statistical approach or a site investigation that is sufficiently detailed to show the uniformity or otherwise of the soil, thus enabling other ultimate bearing capacities of other piles to be estimated. For practical purposes the ultimate bearing capacity may be taken as that load which causes the head of the pile to settle 10% of the pile diameter. According to Price, Malkin and Knill[8], random pile testing does not lend itself to Coal Measures strata due to the rapid variations in lithology which are likely to be met with. In such instances it is necessary to test either the weakest member of the sequence or that rock unit which has been chosen as the foundation level.

1.5. (top) Plate bearing test and (bottom) pile loading test (courtesy D. A. Richardson)

A site investigation must be concluded by a report embodying the findings. This should contain geological plans of the site with accompanying sections, thereby conveying a three-dimensional picture of the sub-surface strata. In this context, if the geology is not too complicated, it may be possible to construct a pin-board, i.e. to erect a pin at each borehole position on a site plan, the individual pins being scale models of the borehole logs. Armed with the details of the foundation conditions the engineer can decide whether or not the ground needs treatment and if so, what type. He can then also decide upon the foundation structure and the various follow-on processes.

Coal Mining

Coal is by far the most important material mined in the UK. It appears to have been worked at least since Roman times. The exploitation of many coalfields dates from the 1100s, the last one to be developed in the UK was the small Kent coalfield, a totally concealed field, which was opened after World War I.

Most of the early workings were at surface outcrops. By the 1300s outcrop workings had largely given way to bell pits and drifting. The scarcity of timber during Elizabethan times led to an increase in the demand for coal and by this time the pillar and stall method of extraction had been evolved. Underground workings were shallow and not extensive, e.g. they might have only penetrated 40 m from the shaft. Indeed, when such limits had been reached, it was usually less costly to abandon a pit and sink another shaft nearby. Evidence of this can still be seen in some rural areas of bygone mining, the surface being peppered with the pock-marks of old shafts. Workings extending 200 m from the shaft were exceptional even at the end of the 1800s, the shaft itself very often being not more than 60 m deep.

With the development of steam power and the use of coal to smelt iron in the 1700s, the demand for coal accelerated. The miner, however, was faced with a number of problems which limited the size and depth of his pit: drainage, ventilation and haulage being three of the most important. Developments such as steam pumps, the Buddle fan, the safety lamp and wire ropes, allowed mines to become more extensive and go to greater depths. In addition, longwall working evolved in the 1700s, probably first originating in the Shropshire coalfield. In this way coal mining gradually developed to become one of the most important industries in the UK, reaching its peak production in 1913.

Subsidence is an inevitable consequence of mining; under some circumstances it may be small or it may be delayed for several years. It can be regarded as the vertical component of ground movement although there is a horizontal component. Subsidence can and does have serious effects on buildings, services and communications; it can be responsible for flooding, can lead to the sterilisation of land, and can call for extensive remedial measures or special constructional design in site development.

Subsidence at the surface reflects the movement which has occurred in the mined-out area. As a consequence small deformation in the workings is associated with insignificant subsidence, whereas total closure of the workings can give rise to severe subsidence at ground level. A number of questions concerning the stability of the surface above mine workings require answers, and these include:

1. Will subsidence occur, and if so, what will be its magnitude?
2. When will it happen and how will it develop?
3. What form will the subsidence take?
4. Is it practical and economic to prevent or reduce its effect?

As already stated, mining methods in which pillars of coal are left as the main support for the overlying rocks have been employed over several centuries. In present-day workings, the pillar supports system can be designed for long-term stability if it is important to protect the surface or to minimise damage to the roof,[9] e.g. the miner can control the pillar dimensions and the percentage extraction. The dimensions of a pillar, which in turn influence its stability, are influenced by the depth of the workings, the seam thickness, the strength of the roof rock (including its fracttture pattern) and local conditions; they must be adequate to prevent the development of creep and thrust. Floor heave and the piling of bulked material, which has collapsed from the roof against pillars, provide a confining effect once an area has been abandoned. The consequences of interaction between beds of widely different properties can be significant, e.g. if a soft layer occurs within a pillar or near the roof or floor, it can have a weakening effect on the more rigid adjacent rock. Materials which display creep properties or are susceptible to deterioration on exposure to air or moisture do not form ideal permanent supports. Extraction ratios often vary between 60 and 70%.

Pillars have to sustain the redistributed weight of the overburden which means that they and the rocks immediately above and below are subjected to compression. Stress concentrations tend to be located at the edge of pillars and the intervening roof beds tend to sag. The effects on ground level are normally insignificant unless the floor is unusually soft. In such cases the pillars may penetrate the floor and give rise to a general lowering of the surface. In addition minor strain and tilt problems occur around the periphery of the basin thereby produced.

Pillars often experience local failures whilst mining is taking place. If a pillar is highly jointed then its margin may fail and fall away under relatively low stress. Such action reduces and ultimately removes the constraint from the core thereby subjecting it to increasing stress. This could lead to pillar failure. Collapse in one pillar can bring about the collapse of others; however, this is not a frequent occurrence in the UK. Collapse of a whole pillar district, however, has been known to happen, e.g. in Coalbrook, South Africa. The Coalbrook colliery disaster occurred at 7.30 p.m. on 21st January, 1960 and lasted for only 5 min[10]. It resulted in the loss of 437 lives and the collapse extended over at least 3 km². The seam in which the collapse took place was 7·6 m thick and at an average depth of 143 m. In such cases the surface suffers substantial rapid subsidence with possible shock damage and often a zone of fracture, severe local strain and tilt damage.

Where old mine workings underlie a site it is necessary to determine the size of the pillars and extraction ratios; very often the pillars were robbed on retreat. At moderate depths the pillar remnants would probably be crushed and the goaf compacted, but at shallow depths, where because of lower pressures the crushing and closure may be variable, this causes foundation problems when large or sensitive structures are to be erected above them. It is therefore necessary to determine the existing load the pillars are carrying and the point at which they will fail. Extraction beneath shallow pillared areas can lead to their collapse and serious subsidence effects at the surface. After a recent site investigation for a hotel in Wideopen, Newcastle-upon-Tyne, it was found that part of the site was underlain by old pillar and stall workings. Moreover, coal was going to be extracted beneath these within the next few years. It was therefore recommended that the hotel complex should be relocated within the site and that the design of the building should be altered.

Even if pillars are relatively stable the surface can be affected by void migration. This can take place over a matter of months or a very long period of years. It has been suggested that void migration will not produce crown holes at the surface where the seam concerned is located at a depth in excess of approximately six times the seam thickness. Nevertheless, voids immediately below the surface create just as awkward a problem. Void migration will be confined below a thick, competent rock unit.

In longwall mining, which is the most important method now used in Europe, the coal is exposed at a face of 30–200 m between two parallel roadways. The roof is supported only in and near the roadways and at the working face. After the coal has been won and loaded, the face supports are advanced leaving the rocks, where coal has been removed, to collapse. Subsidence at the surface more or less follows the advance of the working face and may be regarded as immediate. The curve of subsidence which precedes a working face first causes surface structures to undergo tension strains, then tilt and finally compression. This differential subsidence can cause substantial damage (Fig. 1.6), the tension strains usually being the most effective in this respect. Although subsidence does not cease entirely when the face stops, only small changes then take place. An important feature of subsidence due to longwall working is the high degree of predictability in marked contrast to other forms of subsidence. Methods of subsidence prediction are given in the Subsidence Engineers' Handbook published by the National Coal Board.

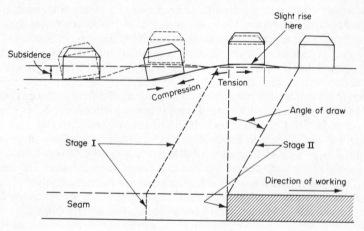

1.6. Wave-like effect of subsidence

A number of factors influence mining subsidence due to longwall extraction. Seam thickness is an obvious one. Maximum vertical subsidence may equal 90% of the thickness of the coal seam extracted. If more than one seam is simultaneously worked beneath the same area then the subsidence effects will be cumulative.

An important factor governing the amount of subsidence is the depth and width of the panel removed. In fact it has been shown that maximum subsidence will begin to occur at a width : depth ratio of 1 : 1·4. This is the critical condition above and below which maximum subsidence is and is not achieved, respectively. The angle of draw defines the outer limit of ground movement; this extends beyond the limits of the goaf. In most British coalfields the angle of draw approximates to 35°. It has been found by field measurements that subsidence is transmitted to the surface almost instantaneously. This, however, does not mean that all the subsidence to which any point at the surface will be subjected will occur immediately. Usually subsidence will continue for a given surface point whilst the face is being worked within the critical area. The amount of subsidence which occurs after the working face has passed out of this critical area is referred to as the *residual* subsidence. This rarely exceeds 5% of the total subsidence at any given point, but such movements may continue for up to two or so years afterwards.

The lithology of the strata between the surface and the coal seam under extraction does not necessarily show a relationship to the amount of subsidence produced. Superficial deposits, however, may allow movements to affect larger areas than otherwise; till sometimes reduces their influence. Some surface rocks, e.g. the Magnesian Limestone, may be badly fractured as a result of mining. Such fractures often have a profound effect on the intensity of differential displacements.

Faults, in particular, tend to be locations where strain is concentrated and unfortunately their exact position at the surface is often difficult to locate. What is more, many coalfields are heavily faulted. If a fault is encountered during seam extraction and its throw is large, then the workings may terminate against the fault; thus permanent strains will be induced at the surface probably accompanied by severe differential subsidence in the zone of influence of the fault. Indeed, a subsidence step may occur at the outcrop of such a fault (Fig. 1.7) sometimes with disastrous effects. The most notable steps occur when the coal is worked beneath the hade of the fault, faces in other positions being much less likely to cause differential movement (Fig. 1.8). Steps are usually down towards the goaf, but if old workings exist then occasionally steps may occur away from the face. The movement is usually vertical and can vary in amount along the fault; it can also be accompanied by horizontal displacement. The size of steps is on average approximately one third of the maximum subsidence which takes place, but this value varies appreciably. The size of step, however, often appears to be consistent where the underground conditions are uniform. The extent of a step is very much limited to the area worked. If coal is only extracted from one side of the fault then the angle of draw can be greatly reduced. By contrast, when workings occur in more than one area the normal angle of draw will probably develop. Once differential movement has occurred further workings in the area cause renewed movement which is sometimes out of all proportion to the thickness and the extent of extraction. The likelihood of movement is not influenced by the depth of the seam but mainly by the position of the face. Because of the significance of faults, the Report on Mining Subsidence prepared by the Institution of Civil Engineers[11] recommends that structures be set back at least 16 m from the line of surface outcrop. On the other hand, a fault can act as a barrier to strains.

When the dip of the strata is less than 20° it has little influence on mining subsidence. The subsidence profile, however, may be shifted towards the deeper end of the worked panel.

Finally the amount of mining subsidence is influenced by filling the goaf with waste material. There are two principal ways in which this can be brought about, namely, by strip packing and by solid stowage. In strip packing the packs are formed by building walls from large stones in the waste material and then filling behind these walls with finer material. This method does not lead to the complete filling of the goaf and in fact is of little value in reducing the amount of subsidence. With solid stowing the goaf is completely filled with waste and can reduce the maximum amount of subsidence by up to 50%. Stowing, however, is costly.

Centuries of coal mining has meant that there are a great number of old shafts left behind. These are now the responsibility of the National Coal Board. Unfortunately the location of many, if not most, of these old shafts is unknown and the location of those which are supposedly known is not always accurate. It was not until 1872 that legislation was introduced making it compulsory to keep plans of mine workings and even then their accuracy cannot be guaranteed.

The oldest and shallowest shafts were timber lined but this was quickly superseded by brick linings. The continuing stability of a shaft lining is obviously

1.7. (top) Subsidence damage near a fault (windows fall out or refuse to open, door frames twist, brickwork cracks, etc.); the dip in the road is due to the fault step. (Bottom) subsidence damage necessitating the shoring up of houses

important, particularly in the uppermost section, for there its collapse generally leads to cratering, i.e. an inrush of surface soils into the shaft to form a depression several times the shaft diameter. A consideration of the long-term stability of shaft linings is therefore required as well as the effects of increased lateral pressure due to the erection of structures nearby.

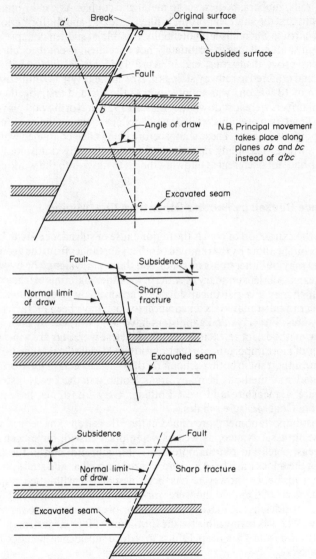

1.8. Influence of faults on subsidence

Filling of shafts in the past was frequently undertaken in a haphazard manner without any regard for the future, e.g. a wooden scaffold may have been constructed just below the surface of a shaft and then topped up with fill. With time the wood decayed to expose an open shaft. Many shafts were filled with unsuitable material and usually no attempt was made to seal them off from the workings, which after the mine was abandoned became waterlogged. In such instances fine fill is likely to flow into the workings with the result that a cavity is produced in the shaft. Again with time the plug above the cavity is denuded until eventually the remnant collapses down the shaft. Usually any material to hand was used to fill a shaft, e.g. rails, timbers, bogies, scrap metal and rubble. As can be imagined this rarely meant that the shaft was properly filled, bridging and the formation of voids usually occurring. Such fills are potentially capable of sudden collapse.

Although shaft collapse is fortunately not a frequent event its occurrence can nevertheless prove disastrous, e.g. in April 1945 an old filled shaft suddenly collapsed and cratered in railway sidings at Abram near Wigan. Immediately after this a train of 13 wagons was shunted across the site and this, together with the engine and driver, were lost down the shaft. In 1956 in Northwood playing fields, Stoke-on-Trent, two previously filled shafts collapsed taking their reinforced concrete cappings with them. These shafts before being filled contained water almost up to the surface. The filling material had obviously proved unsuitable and the mouthings had not been sealed allowing the fill to move into the workings.

Subsidence Caused by Factors Other than Coal Mining

Although the extraction of coal is the major cause of subsidence in the UK, it can be and is brought about by the removal of other substances from the ground. These substances may be either fluids or solids and their removal may be by artificial or natural means. A brief summary of some of the other materials extracted from the ground which may give rise to subsidence is accordingly given below.

A number of other materials, in addition to coal, have been extracted from the Coal Measures. Fireclays and gannisters have been worked, by both quarrying and mining methods, for refractory purposes. These workings are widespread but gannister is of more importance in the Lower than Middle Coal Measures whereas fireclays are important in both, particularly the latter. Fireclays have been worked by pillar and stall methods in many areas, notably in the Leeds district, south Staffordshire, Lancashire and west Lothian. Clay ironstones have also been worked from Coal Measures shales.

The most important metal ore mined in the UK is iron. The sedimentary iron ores of the Jurassic system, as well as being open-casted, have been mined in several areas, chiefly in Northamptonshire, Lincolnshire and in the Cleveland hills. Again the mines have been and are worked on a pillar and stall system. Little significant subsidence, however, has been associated with such mining. The hematite deposits of Cumberland were worked by total extraction methods which gave rise to subsidence, e.g. that due to extraction of ore from the Hodbarrow mine, Millom, by 1922 was responsible for the formation of four large depressions each about 400 m wide and 15 m deep. Efforts to reduce subsidence included hydraulic stowage of sand.

Many other metals have been worked in the UK, notably lead and zinc in the Pennine orefield and the Lake District, and tin and copper in Cornwall. When these are vein deposits the workings follow the vein and are consequently narrow. In addition the host rock is invariably competent, e.g. in the Pennine orefield it is generally the Carboniferous Limestone. Such workings do not as a rule produce subsidence problems. Old mine shafts, however, do constitute a problem. Although dissemination deposits require a greater amount of extraction these again generally do not give rise to significant subsidence. Gangue minerals such as fluorspar, calcite and barytes have been and are also worked in the Pennine orefield. As far as subsidence is concerned very much the same applies to them as it does to their associated metal ores.

Limestone is a carbonate rock which is subject to solution attack which results in the formation of sinkholes (Fig. 1.9) and ultimately subterranean caverns linked together by an integrated system of underground passages. Karstic features are only produced in thick sequences of limestone. Solution features are also

1.9. Gaping Ghyll, a pot-hole which descends some 150 m in limestone of Carboniferous age in Yorkshire

developed in dolomite. Chalk, although subject to solution, because of its relative softness, tends not to allow the formation of subterranean solution features. Rapid subsidence can take place due to the collapse of holes and cavities within limestone which has been subjected to prolonged solution (the time involved being measured on a geological, not human, scale), this occurring when the roof rocks are no longer thick enough to support themselves. Extensive collapse features have in past geological times been produced, witness the sand pits between Brassington and Friden, Derbyshire. Here sand pits, very often over 100 m across, were produced in Tertiary times by the collapse of the roof rocks into caverns in limestone. At this time Triassic rocks covered the Carboniferous Limestone. Accordingly the Bunter Sandstone was the parent material of these sand deposits. It must be emphasised that the solution of limestone is a very slow process; contemporary solution is therefore very rarely the cause of collapse. Nevertheless solution may be accelerated by man-made changes in the ground-water conditions or by a change in the character of the surface water that drains into limestone.

Differential settlement under loading has occurred in limestone which at the ground level appeared competent, but immediately beneath the surface consisted of long, narrow pinnacles of rock separated by solution channels occupied by broken limestone rubble in a clay matrix.

Limestone is occasionally mined, but because of its competence does not give rise to subsidence of any significance, e.g. the Hopton Wood Limestone is worked at Middleton-by-Wirksworth, Derbyshire (Fig. 1.10). The pillar and stall workings are on two levels, the individual pillars being approximately 148 m², by 5·2 m in height. They support over 137 m of limestone overburden and the amount of subsidence has been negligible. The Bath Stone (Great Oolite) has been worked in Wiltshire and Somerset, e.g. within the Box–Corsham district of Wiltshire there are 96·6 km² of pillar and stall workings. Formerly, the extraction ratio reached 85%, but it has been reduced to 65%.

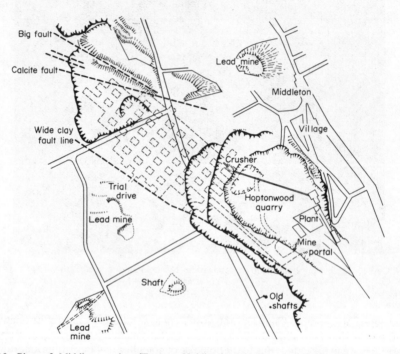

1.10. Plan of Middleton mine (Tarmac Holdings), Middleton-by-Wirksworth, Derbyshire (courtesy *Cement, Lime and Gravel*)

Shallow mine workings occur in the Chalk, particularly in East Anglia and Kent. Some of these were made by Palaeolithic man in his quest for flint. These are usually similar to bell-pits; they may be about 3–6 m in height, and have small tunnels, following the flints, running from the main chamber. Chalk has also been worked at later times for building and agricultural purposes.

A notable example of subsidence in chalk occurred in Bury St. Edmunds in July 1966 (Fig. 1.11). The subsidence took the form of crown holes and happened in Jacqueline Close, with the result that the houses were eventually declared unsafe and therefore abandoned. As far as the workings are concerned the pavement is usually between 15 and 18 m below the surface and the galleries 2–3·5 m high. Although the ground was potentially unstable, in that crown holes were likely to occur, the process was accelerated by site development which meant a change in the subsurface drainage regime. It appears that soakaway water led to extra solutioning and the rapid decay of the roofs above galleries.

1.11. Subsidence in the form of crown holes in the Chalk, Jacqueline Close, Bury St. Edmunds (courtesy *East Anglian Daily Times*)

Gypsum is more readily soluble than limestone, e.g. 2100 p.p.m. can be dissolved in non-saline waters as compared with 400 p.p.m. Caverns can therefore develop in thick beds of gypsum more rapidly than they can in limestone; indeed they have been known to form in the USA within a matter of a few years where thick beds of gypsum have been located below dams. Extensive surface cracking and subsidence have been attributed to the collapse of cavernous gypsum in states like Oklahoma and New Mexico. The problem is accentuated by the fact that gypsum is weaker than limestone and therefore collapses more readily. The solution of gypsum gives rise to sulphate-bearing waters and these may attack concrete.

A problem associated with anhydrite which has also occurred in the USA is uplift. This presumably takes place when anhydrite is hydrated to form gypsum; in so doing there is a volume increase of between 30 and 58% which exerts pressures which have been variously estimated between 2 and 69 MN/m^2. It is thought that no great length of time is required to bring about such hydration. When it occurs at shallow depths it causes an expansion, but the process then appears to be gradual and is usually accompanied by the removal of gypsum in solution. At greater depths anhydrite is effectively confined during the process. This results in a gradual build-up of pressure and the stress is finally liberated as an explosive force. Such uplifts have invariably taken place beneath streams or standing bodies of water, artificial or natural. It would appear that a body of water is necessary to provide a constant supply for the hydration process, percolation being via cracks and fissures. Examples are known of ground being elevated by about 6 m, along 300 or so metres of stream length. Such rapid explosive movement causes strata to fold, buckle and shear which then further facilitates access of water into the ground.

Gypsum, which occurs in the Permian and Triassic systems, has been mined in various parts of the UK, e.g. in the Lake District, Durham, Nottinghamshire and Staffordshire. The pillar and stall method of extraction was used. Most of the beds which were worked did not exceed 5 m in thickness and the extraction ratio was not generally more than 75%. Such mines are not generally extensive, at the most extending over approximately 4×10^5 m², but some were located at shallow depths. They have not given rise to any significant subsidence problems.

Salt is even more soluble than gypsum and the evidence of slumping, brecciation and collapse structures in rocks which overlie saliferous strata bear witness to the fact that salt has gone into solution in past geological times. It is generally believed, however, that in areas underlain by saliferous beds measurable surface subsidence is unlikely to occur except where salt is being extracted. Perhaps this is because equilibrium has been attained between the supply of unsaturated ground-water and the salt available for solution. Exceptionally, cases have been recorded of rapid subsidence, e.g. the 'Meade salt sink' in Kansas was explained by Johnson[12] as due to solution of deep salt beds. This area of water, about 60 m in diameter, occurred as a result of subsidence in March 1879. At the same time, 64 km to the south west, the railway station at Rosel and several buildings disappeared due to the sudden appearance of a sinkhole.

The occurrence of salt springs in Cheshire has been quoted as evidence of natural solution and it is thought that the Cheshire meres were formed as a result of the subsidence which thereby occurred, but such subsidence usually is a very slow process operating over large areas. Localised solution is presumed to develop collecting channels along the upper surface of salt beds. The incompetence of the overlying Keuper Marl inhibits the formation of large voids, roof collapse no doubt taking place at more or less the same time as salt removal. This action eventually gives rise to linear subsidence depressions or brine runs at ground level.

Classic examples of subsidence due to salt working have occurred in Cheshire where salt has been extracted for over 300 years. There the salt occurs in two principal beds, the upper one of which was worked by pillar and stall methods as early as the seventeenth century. The lower has been worked since the nineteenth century, predominantly by brine pumping. Collapse of pillars in the old mines of the upper bed has caused funnel shaped hollows of various diameters and depths to occur at the surface. But the most disastrous subsidences occurred in the second half of the nineteenth and early twentieth centuries as a result of bastard brine pumping, i.e. extracting brine from old mine workings and thereby causing their collapse (Fig. 1.12). Natural brine pumping, which still continues, has produced surface troughs of various sizes and shapes, depending on the amount of pumping and the competency of the overlying rocks. As the brine runs extend so the subsidence and existing depressions are enlarged and deepened. It is usually impossible to predict the extent and amount of future subsidence since the width and height of the area undergoing solution is almost always unknown. However, the modern system of producing brine by controlled solution is virtually subsidence-free. Salt has also been extracted in Lancashire, Staffordshire, Worcestershire and Teesside.

An interesting example of past mining which was not realised to be as extensive occurs in the Upper Greensand near Merstham, Surrey. Here the rock was worked for hearthstone by a crude pillar and stall method, the extraction ratio varying between 65 and 75%. The workings occur around and beneath the M23/M25 interchange and therefore required extensive grouting.

Subsidence may be caused as a result of plastic flowage of subsurface rocks. Cambering has occurred in Northamptonshire. Cambers are structures in outcropping or near surface strata which have been let down along the sides of valleys as a consequence of plastic flow towards the valley of underlying material, in this case Liassic clay. Clay is squeezed from the base of scarps to form valley bulges. Plastic flow may also occur in salt or gypsum.

1.12. Disastrous effect of subsidence in Northwich, Cheshire, 1900 (from A. F. Calvert, *Salt in Cheshire* (1915))

In a very few localities the creation of subsurface cavities by mechanical erosion of sediment has been recognised as the reason why the ground above subsequently collapsed. The sediment must be easily erodible and transportable and be overlain by material competent enough to maintain a temporary roof support. Silts and sands may fulfil such a role. Terzaghi[13] diagnosed an apparently mysterious subsidence in Memphis, Tennessee as having been caused by the subsurface erosion of quartz sand.

The removal of fluids from sediments reduces their pore pressures and thereby causes them to consolidate. This in turn leads to subsidence. Peat is the most compressible of materials, and is highly porous so that its water content may range up to 2000%. Its specific gravity is low, usually about 1·0 and it has little strength. Drainage of peat leads invariably to subsidence, the Fenlands providing a classic example. There peat has been drained for over 400 years. In some parts of the Fens the thickness of peat has almost been halved as a result, e.g. in the years between 1848 and 1932 a total subsidence of 2·7 m was recorded by the Holme Post, the original thickness of peat being 6·7 m (Fig. 1.13).

It is estimated that drainage of the Dutch polderlands caused 30–35% consolidation in clays and 20% in silts over a period of about 100 years. Consolidation in sands is negligible.

Abstraction of water from the Chalk over the past 150 years has caused subsidence in some areas of London in excess of 0·3 m. In 1820 the artesian head in the Chalk was approximately +9·1 m o.d., but by 1936 this had declined in some places to −90 m o.d. The decline in artesian head has been accompanied by underdrainage in the London Clay. Between 1865 and 1931 subsidence averaged between 60 and 180 mm throughout much of London (Fig. 1.14).

Subsidence in Mexico City due to the abstraction of water from a sand-gravel aquifer has been so impressive that it is now an often-quoted classic example. The aquifer extends under the city from an approximate depth of 50 m below ground level to well below 500 m. Water has been abstracted for about 100 years, but at a generally increasing rate. By 1959 most of the old city had suffered at least 4 m of subsidence, and in the north east part, as much as 7·5 m had been recorded.

1.13. The Holme Post, a cast-iron pillar erected in 1851 on the south west edge of Whittlesey Mere. It replaced the wooden posts which were erected in 1848 to indicate peat shrinkage caused by drainage. The post was driven 7 m through peat into clay until its top was flush with the ground. Within 10 years ground level had fallen 1·5 m through shrinkage. A second post was erected in 1957 with its top at the same level as that of the original post (right-hand side). Between 1850 and 1970 the ground fell some 4 m

A spectacular and costly subsidence occurred at the Wilmington oilfield in the harbour area of Los Angeles and Long Beach, California. This was first noticed in 1940 and by 1962 the subsidence had increased to about 8 m at its centre (the greatest subsidence due to fluid withdrawal known to have occurred in the world), and included an area of about 64 km^2 that had subsided 0·6 m or more (Fig. 1.15). Incidentally, many oilfields of similar character and exploitation to that of Wilmington have experienced subsidences of less than 1 m. The problem of subsidence was acute since much of the area affected was only 1·5–3·0 m above sea level and was highly industrialised. In 1947 subsidence was occurring at the rate of 0·3 m/year; this increased to 0·7 m/year in 1951 when maximum extraction was reached, i.e. 140 000 barrels per day. After this, production declined, so did the rate of subsidence. Because of the seriousness of subsidence remedial action was taken in 1958 by repressuring the oil zones by injecting water. By 1962 this had brought subsidence to a halt in much of the field. The area is underlain by about 1800 m of Miocene to Recent sediments consisting largely of sands, siltstones and shales. The reason for the subsidence is given as the decline in fluid pressure in the oil zones due to the removal of oil, gas and water, this inducing consolidation.

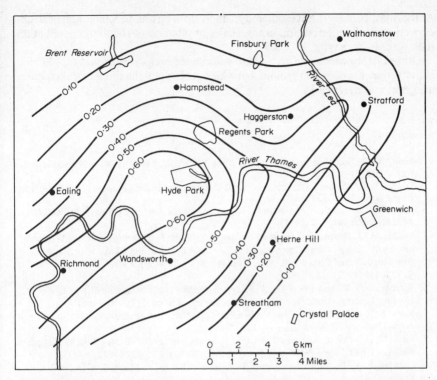

1.14. Lines of equal subsidence (1865–1931) due to abstraction of water from the Chalk beneath London; the contours are in tenths of a foot (after Wilson G. and Grace, H., 'The Settlement of London due to Underdrainage of the London Clay', *J. Inst. Civ. Engng,* **19,** 100–127 (1942))

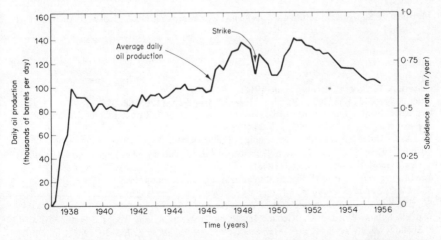

1.15. Oil production and subsidence rate at Wilmington oil field, California (after Steinbrugge, K. V. and Bush, V. R., *Subsidence in Long Beach–Terminal Island–Wilmington, California,* Pacific Fire Rating Bureau, San Francisco (1958))

Authorities, however, have differed in their opinions as to which material underwent consolidation: the oil sands, shales or siltstones. Probably consolidation took place in each type.

Other factors which can give rise to subsidence such as earthquakes, volcanic activity, thawing of frozen ground, but which are beyond the scope of this chapter, are given in Reference 14.

REFERENCES

1. Scott, A. C., 'Locating and Filling Old Mine Workings', *Civ. Eng. Pub. Works Rev.,* **52**, 1007–1011 (1957)
2. Deere, D. U., 'Technical Description of Rock Cores for Engineering Purposes', *Rock Mech. and Engng Geol.,* **1**. 18–22 (1963)
3. Franklin, J. A., Broch, E. and Walton, G., 'Logging the Mechanical Character of Rock', *Trans. Inst. Min. Met.,* **81**, 1–9 (1971)
4. Fookes, P. G., Dearman, W. R. and Franklin, J. A., 'Some Engineering Aspects of Weathering with Field Examples from Dartmoor and Elsewhere', *Quart. J. Eng. Geol.,* **3**, 1–24 (1972)
5. Franklin, J. A. and Chandra, R., 'The Slake Durability Test', *Int. J. Rock Mech. and Min. Sci.,* **9**, 325–341 (1972)
6. Raybould, D. R. and Price, D. G., 'The Use of the Proton Magnetometer in Engineering Geology Investigations', *Proc. 1st Cong. Rock Mech.,* Lisbon, **1**, 11–14 (1966)
7. Early, K. R. and Dyer, K. R., 'The Use of a Resistivity Survey on a Foundation Site Underlain by Karst Dolomite', *Geotechnique,* **14**, 341–348 (1964)
8. Price, D. G., Malkin, A. B. and Knill, J. L., 'Foundations of Multi-Storey Blocks with Special Reference to Old Mine Workings', *Quart. J. Eng. Geol.,* **1**, 271–322 (1969)
9. Orchard, R. J., 'Partial Extraction and Subsidence', *Min. Engr.,* **123**, 417–427 (1964)
10. Bryan, A., Bryan, J. G. and Fouche, J., 'Some Problems of Strata Control in Pillar Workings', *Min. Engr.,* **123**, 238–266 (1964)
11. *Report on Mining Subsidence,* Inst. Civ. Eng., London (1959)
12. Johnson, W. D., 'The High Plains and Their Utilisation', U.S. Geol. Surv., 21st Annual Rept., part 4, 601–741 (1901)
13. Terzaghi, K., 'Earth Slips and Subsidence for Underground Erosion', *Eng. News Record,* **107**, 90–92 (1931)
14. Allen, A. S., *Geologic Settings of Subsidence,* Reviews in Eng. Geol., Geol. Soc. Am., 305–341 (1963)

BIBLIOGRAPHY

Calvert, A. F., *Salt in Cheshire,* Spon (1915)
Dean, J. W., 'Old Mine Shafts and Their Hazards', *Min. Engr.,* **126**, 368–377 (1967)
Geological Society Engineering Group, Working Party on the Logging of Cores for Engineering Purposes, *Quart. J. Eng. Geol.,* **3**, 1–24 (1970)
Lee, A. J., 'The Effect of Faulting on Mining Subsidence', *Min. Engr.,* **125**, 735–743 (1966)
Poland, J. F. and Davis, G. H., 'Land Subsidence Due to Withdrawal of Fluids', *Reviews in Eng. Geol.,* Geol. Soc. Am., 190–269 (1963)
Site Investigations, CP 2001, British Standards Institution, London (1957)
Subsidence Engineers Handbook, National Coal Board, London (1966)
Wardell, K., 'Mining Subsidence', *Trans. Inst. Min. Surv.,* **34**, 53–70 (1954)
Wardell, K. and Wood, J. C., 'Ground Instability Problems Arising from the Presence of Old, Shallow, Mine Workings', *Proc. Mid. Soc. Soil Mech. Found. Eng.,* **7**, 5–30 (1966)

Chapter 2

The Character of the Coal Measures

The term *Coal Measures* is given to those rocks which occur in the Carboniferous period above the Millstone Grit, or Namurian as it is now known. The Carboniferous period is divided into lower and upper divisions, the Namurian and Coal Measures being in the latter. As can be inferred from the name, this sequence of rocks contains workable coals. Workable coals with associated rock types, however, are found in other parts of the Carboniferous of the UK, e.g. in the Scremerston Coal Group and the Limestone Group of the Lower Carboniferous of Northumberland, and in the Edge Coal Group of Scotland. Furthermore, the uppermost strata of the Coal Measures, known as the Stephanian, are barren, i.e. they do not contain workable coals. That part of the Coal Measures which does contain workable coals is referred to as the Westphalian. Coal occurs throughout the world in every system from the Devonian onwards. This is because of the evolution of land plants in Devonian times.

Coal has been and is of enormous economic importance. It has been worked since Roman times and many coalfields have been exploited for centuries. More importantly, coal played a major role in the industrial revolution in Britain and led to the establishment of major industrial centres on coalfields. Indeed, with the notable exception of London, the major centres of population in the UK today are located on or near coalfields. The present redevelopment of these urban areas must give due regard to past and present mining activities if trouble is to be avoided.

The Coal-Measure Environment

Coal is an organic deposit which occurs in association with other sedimentary rocks such as mudstones, shales and sandstones. Its microscopic study reveals that it is composed of different types of macerated plant tissue. It would therefore appear that a prerequisite for the development of a coal seam was a prolific vegetative cover. What is more, true coal seams have certain characteristics which suggest that the vegetable debris accumulated *in situ* rather than being carried into the basin of deposition. The forests which provided the vegetable debris were extensive; their elevation was very near sea level and they grew in badly drained swamplands.

In order for a sufficient quantity of vegetable matter to accumulate (it is thought that between 10 and 15 m of peat would produce 1 m of coal), subsidence would have to more or less keep pace with growth of organic debris. If accumulation took place faster than subsidence, then the organic deposit would build up and thereby tend to drain itself. The consequent loss of acidic water would lead to a more rapid decay of the organic deposit leading ultimately to its disappearance. In order that the plant debris be preserved, bacterial decay and oxidation must be retarded; stagnant, deoxygenated, waterlogged conditions providing the right environment. In coal-forest swamps the lack of water movement meant that it was poorly aerated and the rotting vegetation no doubt meant that these waters were acidified. After the organic deposit was formed it had to be buried. This was brought about by more rapid subsidence and the simultaneous deposition of muds or sands over the plant material. Rapid and complete burial would reduce or possibly arrest the rate at which the plant material was decaying. In order for another coal seam to be produced, these sediments would have to build up to a level where plant growth could be re-established. The general sequence of events was that a deposit of plant material was first inundated and covered with muds, this being brought about by more rapid subsidence. The arrival of coarser sediments may indicate the appearance of an advancing delta front or be due to increased river speed, the latter being influenced by tectonic control. With the deposition of sands, the water level was reduced and eventually became shallow enough for plant growth to occur again.

The change from plant debris to peat is a biochemical transformation, but that from peat to coal is a dynamo-chemical process which is only brought about by burial. The process of coalification is reflected in the rank of coals. 'Rank' refers to the chemical development of coals, more particularly to the increase in the proportion of carbon with decreasing content of oxygen and hydrogen. With increasing rank, peat changes to lignite, lignite to bituminous coal and the latter to anthracite.

A characteristic feature of Coal Measure sedimentation is its rhythmic deposition with the consequent development of cyclothemic stratal sequences (Fig. 2.1). A cyclothem has been defined as a multiple repetition of beds of different lithology which are recognisably similar in internal sequence, showing usually minor variations both in thickness and precise sequence of components. Some rock types may be absent from individual cyclothems. In the UK the simple cyclothemic unit in the Coal Measures usually consists of:

$$
\text{One cyclothem}
\begin{cases}
\text{Coal} \\
\text{Seatearth} \\
\text{Sandstone} \\
\text{Non-marine shale or mudstone} \\
\text{Marine band} \\
\text{Coal}
\end{cases}
$$

Little reliance, however, may be placed on the standard pattern in a particular coalfield, as far as a means of predicting rock types at various levels within the sequence is concerned. Marine bands are generally thin and indeed in many units are absent; coals may also be absent. Individual cyclothems vary in thickness usually ranging between a few metres up to a few tens of metres. Fine-grained sediments form the major part of a cyclothem, the ratio of shale, mudstone and siltstone to sandstone being about $3:1$. It follows that the thickest members of a cyclothemic unit are generally the mudstones and shales, but massive bedded sandstones may be important locally. These, however, are often lenticular and laterally impersistent. The coal in a given thickness of coal-bearing strata will usually form from 2 to 5% of the total thickness, inclusive of thin, unworkable seams. The coal seams themselves vary in thickness from mere films to a couple of metres or so thick. It has been suggested that the irregular and repeated subsidence necessary to explain cyclothemic phenomena could have been brought about by crustal movements or glacial control.

Most coal seams have a composite character. At the bottom the coal is often softer and is sometimes simply referred to as *bottom* coal. In the centre of the seam, bright coal is often of most importance whilst dull coal may predominate in the upper part of the seam. This change in the nature of a coal seam may be explained by a

B.H.	Depth (m)	Core recovery	Description of strata
		—	Asphalt path and filling
	0·46		Stiff grey clay
	0·76	70%	
			Alternating hard black mudstones, shales and soft clays
	1·5		
		35%	Firm grey contorted sandy clay with brown iron staining
	2·9		
		30%	Hard grey mudstone and firm grey shaley clay
	4·4		
		Wash	Firm grey-brown clay or soft mudstone
	5·6		
		25%	Very hard grey sandstone with minor lenticular argillaceous and carbonaceous partings
	7·3		
		40%	Hard grey shaley mudstone with sandstone partings, minor iron stained fissures
	8·5		
		60%	Hard grey sandstone with fine carbonaceous partings and a fine shaley mudstone band
	10·5		
		70%	Medium-hard dark-grey shaley arenaceous mudstone
	11·3		
		75%	Hard grey sandstone with fine carbonaceous partings becoming micaceous
	12·3		
		Zero	Cavity (old workings?)
	13·6		
		Zero	Carbonaceous material (roof fall?)
		50%	
	16·0		
	16·2	Traces	Stiff grey fireclay

2.1. Typical borehole log of Coal Measures strata

change in the type of plant material deposited and changing drainage conditions. A seam may have a clean-cut top against the roof, or dirt partings may make their appearance. Occasionally, cannel coal may occur at the top of a seam. Such features may aid the identification of individual coals.

Most coals can be broken into blocks which have three faces approximately at right angles. These surfaces, along which breakage takes place, are termed *cleat*. The cleat direction is usually pretty constant and frequently influenced the direction of underground working, for it was easier to hew across the cleat than along it. Within a single seam, the cleat is best developed in the bright coals and less prominent in the dull. Cleat partings may be filmed with mineral matter, commonly calcite, ankerite and pyrites. Cleat is not developed in anthracite.

In contrast to the slow, quiet, stable deposition of plant debris to form coals, the interseam material was presumably laid down relatively rapidly, sometimes in turbulent conditions. Contemporary adjustment to depositional instability, especially during dewatering, manifested itself in the form of micro-faulting and thrusting, and by the injection of sandstone dykes and sills. Load casts and convolute lamination are frequently present in coal-bearing sediments, the latter is typically developed in interlaminated sand–silt associations. Seismic activity appears to have occurred at various times throughout the history of these subsiding basins, witness the presence of distorted, folded and broken shales, mudstones, siltstones and sandstones.

Differential subsidence caused coal seams to split, muds or sands accumulating in areas of more rapid sinking. Such sedimentation could build up to the level of peat accumulation so that deposition of peat again became continuous. A good example of coal splitting is found in south Staffordshire where the Ten Yards Coal, when traced some 8 km to the north, splits into a dozen or more seams, the partings between which total some 150 m. Those areas where coal seams are replaced either totally or partially by clastic sediments, often fairly coarse grained, are known as *washouts*. Coal swamps are frequently imagined to have developed in deltaic areas. Washouts are assumed to have been formed in distributary channels, the distributaries depositing sands and muds whilst peat accumulated along their banks. They could also have been formed after the deposition of peat, newly established distributary streams eroding into and through the peat and leaving sand or mud in its place. Washouts are generally lens-like in shape and may affect more than one seam. Those which affect several seams are usually encountered when the seams are relatively close together.

Inter-Seam Rocks

Argillaceous rocks account for about three quarters of the thickness of a sequence of coal-bearing strata. Their occurrence is often widespread and uniform and they may be traceable over large areas of a coalfield. The individual beds may be over 30 m in thickness. These argillaceous rocks are usually some shade between black and pale grey, although blue-grey is a common colour, and brown and red mudstones and shales occur, particularly in the upper part of the Coal Measures. There are several types of black shales, e.g. carbonaceous shales are soft, finely laminated and feel rather soapy. As their name suggests they contain plant remains. Marine shales may also be black in colour but these are not nearly as frequent in occurrence as non-marine shales.

There is no sharp distinction between mudstones and shales, one grading into the other. Shales, however, are characterised by their lamination, which in some cases may be very thin, as in 'paper' shales, whilst mudstones are relatively massive and may break along joints or with an irregular, conchoidal fracture (Fig. 2.2). Clays, like mudstones, are relatively massive, but they lack induration.

2.2. Typical sequence of mudstones and shales with some thin sandstones; Coal Measures, south Wales

The principal components of shales, mudstones and clays are clay minerals. Finely comminuted quartz, frequently less than 0·001 mm in diameter, is the second most important mineral, e.g. many mudstones may contain somewhat less than 20% of quartz. Feldspars, micas and carbonaceous debris may be present in varying small amounts. The plasticity of a typical clay from the Coal Measures is due to the presence of a sufficient quantity of clay minerals and of colloidal material with their marked affinity for water. Some marine shales may be impregnated with carbonate material, which makes them appreciably stronger, and they may grade into impure limestones. Iron pyrites is also found in argillaceous rocks, especially the shales. This is a relatively unstable mineral and its breakdown in the presence of water may lead to the formation of sulphuric acid. Relatively low pH values may be recorded in the small, stagnant pools that occur on waste tips.

Seatearths are almost invariably found beneath coal seams (Fig. 2.3). They may occur at other levels in a sequence, but generally they can be regarded as marking the end of a rhythmic cycle. The character of a seatearth depends on the type of deposits which were laid down immediately before the establishment of plant growth. If they were muds then a fireclay underlies the coal; on the other hand if they were silts and sands then it is a gannister. Seatearths can be regarded as fossil soils and as such are characterised by the presence of fossilised rootlets. These rootlets extended downwards and tended to destroy the lamination and bedding of the seatearth. Seatearths are presumed to have undergone significant leaching, which accounts for their small quantities of magnesia, lime and alkalis. It also accounts for the presence of kaolinite in fireclays and the high silica content of gannisters. Fireclays may contain 50–75% silica and 15–35% alumina, compared with gannister which contains 95–98% silica and less than 1% of alumina. A typical fireclay is pale grey and consists of clay minerals with fine quartz grains.

2.3. Coal seam with fireclay beneath

Fireclays which contain some iron are yellow at outcrop; the presence of organic material produces darker shades. Fireclays with a low quartz content are typically highly slickensided and break easily along randomly orientated listric surfaces. A gannister is a fine-grained quartzose rock, often the quartz grains are angular and of rather uniform size, frequently between 0·05 and 0·025 mm diameter. Thus many gannisters are pure siltstones and because they are usually well cemented with silica they are hard and strong. They are usually light grey in colour although the presence of carbonaceous matter may again give darker shades. Individual beds of gannister only infrequently exceed 2 m in thickness.

Clay ironstone may be found in the mudstones and shales of the Coal Measures. In fact in former times they were an important source of iron ore although today none are worked. Most clay ironstones are nodular and are principally composed of siderite and chalybite with a variable amount of clay. The iron content is generally between 25 and 30%. In the north Staffordshire and Scottish coalfields blackband ironstone may contain up to 20% carbonaceous matter. The ironstone nodules vary in size, some being less than a pea whilst others exceed 0·3 m in length. The larger ones are elongate, ovoid in shape, their long axis being parallel to the bedding direction, and they are often septarian. Nodules frequently occur in a recognisable layer and in some instances they may be replaced by a continuous band of ironstone. The shales in which they occur are typically iron stained and brittle.

Siltstones are common members of the inter-seam rocks. They grade imperceptibly into rocks of clay grade. However the gritty nature of a siltstone may be felt or indicated by scraping with a knife, thus allowing its distinction to be made. Some are massive although many are laminated. The individual laminae may be picked out by the presence of darker layers which contain carbonaceous matter and/or mica. They vary in thickness from less than 1 mm up to a few millimetres. Micro-cross bedding is frequently present and in some siltstones the lamination may be convoluted. Siltstones have a high quartz content with a predominantly siliceous cement; they are therefore grey in colour and are hard, tough rocks. Frequently siltstones are interbedded with shales or fine-grained sandstones, the siltstones occurring as thin ribs.

Sandstones are the most inconsistent members of Coal Measures cyclothems. They are generally lenticular in shape and vary rapidly in thickness. Most of the sandstones found in the productive part of the Coal Measures are greyish in colour, the shade depending to some extent on the amount of carbonaceous material they contain. On weathering, the sandstones tend to become yellow or brown. Some of the sandstones are composed almost entirely of quartz and usually have a siliceous cement. These are white or pale grey in colour and are amongst the hardest rocks of the Coal Measures. Other minerals found in sandstones include feldspars, which show varying degrees of alteration, and micas. Some sandstones may have a carbonate-cement or argillaceous matrix, or an admixture of both. Iron oxides may also act as a cementing material. The majority of the sandstones, however, are subgreywackes with a lesser number of quartz arenites, subquartz arenites and subarkoses.

Most of the inter-seam sandstones are fine-to-medium grained, although occasional coarse varieties do occur. Pebbles are often found in the latter types, particularly near the base of a bed. Many coarse-grained sandstones often contain plant remains and coaly debris. These coarse-grained sandstones may merge into conglomerates and breccias. Generally speaking, Coal Measures sandstones may be subdivided into two groups, the thick-bedded massive or coarsely cross-bedded types on the one hand (Fig. 2.4), and the fine-grained, thinner and less-persistent

2.4. Typical massive Coal Measures sandstone showing bedding planes and joint surfaces. Note the plumose structure on the joint surface in the left, centre, of the exposure

types on the other. The latter type may exhibit cross-bedding on a macro- or micro-scale, they are frequently laminated, notably the micaceous varieties, and ripple marks may also be present. Some of these sandstones may be poorly sorted whereas the quartz arenites are generally moderately-to-well sorted. Some of the sandstones, particularly the coarser varieties, are quite porous and act as reservoirs for underground water.

Many of the rocks of the Coal Measures have been worked besides coal. Both gannister and fireclays are worked for refractory purposes and have been either quarried or mined. As mentioned above, a gannister is a seatearth with a high silica content, the bulk of all the true gannisters occurring in the Lower Coal Measures. Gannisters are principally used for making silica bricks and for moulding sands. Fireclays tend to replace gannisters as seatearths in the Middle Coal Measures. The best fireclays are soft, greyish clays of moderate plasticity and are used to make firebricks. Many between coal shales, the so-called bastard fireclays, although not refractory enough for firebricks, have been worked for sanitary ware. Shales, mudstones and clays are extensively worked for brick making, indeed some of the best quality engineering bricks are made from these argillaceous rocks. Ironstones were formerly quarried and mined on an extensive scale, although none are exploited at the present time. Many sandstones have also been quarried and mined for flags, building stones, moulding sands and grindstones. Consequently the engineer who has to carry out a site investigation in an area of Coal Measures should bear in mind that problems may arise not only due to the extraction of coal, but because of the exploitation of other materials, whether in the past or present. Past exploitation will often provide more headaches than that going on at present.

Strength of Coal Measures Strata

The behaviour of a rock under stress is influenced by several external factors such as pressure, temperature conditions, pore solutions and time. It has been frequently shown that the ultimate compressive strength, i.e. the point immediately before rupture, is many times increased by high confining pressures. At such pressures incipient fractures in the rock are closed and the total flow of material without rupture may be indefinitely increased with increasing confining pressure. High temperatures tend to lower rock strength and bring about plastic deformation; they become increaaingly important with increasing depth. As many rocks become more 'plastic' under the influence of high confining pressures and increasing temperatures, the angle between the conjugate shear fractures widens about the axis of maximum stress. In some sandstones, however, there is virtually no change in the angle of shear up to several hundred meganewtons per square metre confining pressure.

The presence of solutions in the pore spaces of a rock causes an increase in strain velocity and lowers its fundamental strength. Colback and Wiid[1] carried out a number of uniaxial and triaxial compression tests at eight different moisture contents on quartzitic sandstone with a porosity of 15%. The tests indicated that the compressive strength under saturated conditions was half what it was under dry conditions. They showed, however, that the coefficient of internal friction was not significantly affected by changes in moisture content. Colback and Wiid[1] therefore tentatively concluded that the reduction in strength witnessed with increasing moisture content was primarily due to a lowering of the tensile strength which is a function of the molecular cohesive strength of the material. Tests showed that the uniaxial compressive strength of quartzitic sandstone was inversely proportional to the surface tension of the different liquids into which it was placed. As the surface free energy of a solid submerged in a liquid is a function of the surface tension of the liquid and since the uniaxial compressive strength is directly related to the uniaxial tensile strength and this to the molecular cohesive strength, it was postulated that the influence of the immersion liquid was to reduce the surface free energy of the rock and hence its strength. The authors therefore concluded that the reduction in strength from the dry to the saturated condition of predominantly quartzitic rocks was a constant which was governed by the reduction of the surface free energy of quartz by the presence of any given liquid.

Numerous tests have been conducted to investigate the change in ultimate strength with time. Furthermore, the study of time-dependent strain or creep in a stressed material constitutes an important part of any rheological investigation. The time–strain pattern exhibited by a wide range of materials subjected to a constant uniaxial stress can be represented diagrammatically as shown in Fig. 2.5. The

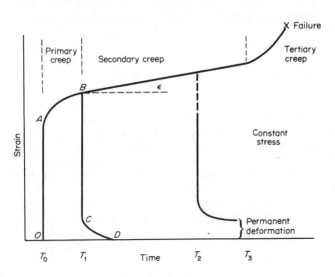

2.5. Theoretical time–strain curve at constant stress (after Price, N. J., *Fault and Joint Development in Brittle and Semi-brittle Rock, Pergamon (1966)*)

instantaneous strain, which takes place when a load is applied, is represented by *OA*. There follows a period of primary or transient creep (*AB*) in which the rate of deformation decreases with time. Primary creep is the elastic effect attributable to intragranular atomic and lattice displacements. If the stress is removed, the specimen recovers; at first this is instantaneous (*BC*), being followed by a time elastic recovery (*CD*). On the other hand, if the loading continues, the sample begins to exhibit secondary or pseudo-viscous creep. This type of creep represents a phase of deformation in which the rate of strain is constant and is due principally to movements which occur on grain boundaries. The deformation is permanent and is proportional to the length of time over which the stress is applied. If the loading is still further continued, then the specimen suffers tertiary creep in which the strain rate accelerates with time and ultimately leads to failure.

The mechanical behaviour of a rock is also influenced by its inherent character, i.e., its mineralogical composition, its texture and the presence of micro-fissures. The varying strength of the minerals of which the rock is composed obviously influence its strength, the proportion of the individual minerals being significant in this context, e.g. it has been shown that the strength of several Coal Measures rocks increases with increasing quartz content, in some cases the relationship being linear. More important is whether or not the constituent minerals have been altered by weathering, for it appears that a very slight degree of weathering can reduce the strength of a rock by a considerable amount. Indeed attempts have been made to define rock quality indices based on the proportions of fresh to altered minerals and the presence of micro-fissures in a specimen. One of the most important facets of texture in relation to rock strength is the manner in which the constituent grains interlock and yet another is the amount of unfilled pore space present. If all minerals are closely interlocked, then the rock has a high strength, whereas poorly interlocking grains and large pore spaces reduce rock strength. Porosity is therefore an important feature of sedimentary rocks and there is a general increase in their compressive strength with decreasing porosity. Indeed Price[2] showed that as far as certain Coal Measures rocks were concerned, for every 1% increase in porosity the completely dry strength decreased by about 4% of the estimated strength at zero porosity. Again, as far as sedimentary rocks are concerned, the amount and, to a lesser extent, the type of cement/matrix is an all-important factor governing their strength. The better the component grains are bound together the stronger the rock; generally siliceous and ferruginous cements give stronger rocks than calcareous cements or a clay matrix. Moreover the more a sedimentary rock is compacted the stronger it is, compaction reducing porosity. As a general statement it can be said that the larger the grain size of a rock the lower its strength.

The ultimate compressive strength of a test specimen is defined as the ratio of the maximum load at failure to the cross-sectional area of the specimen before the test. Failure is considered to have occurred when no further load can be supported. The uniaxial compressive strength of a rock is one of the simplest measures of strength and although its application is limited it does allow comparison to be made between rocks. It also affords some indication of rock behaviour under more complex stress systems. The behaviour of a rock under uniaxial compression is influenced by the test conditions. The most important test variables are the shape of the specimen, the dimensions of the specimen, its end conditions, and the rate of loading. As far as shape is concerned, cylindrical samples are usually tested; cubic samples will give higher strengths. When cylindrical samples are tested, the ideal length : diameter ratio is 2–2·5 : 1. This gives a reasonably good distribution of stress throughout the specimen and eliminates the confining effects of the platens. The ends of the specimen should be parallel and at right angles to the central axis. This means that they should be lapped and polished after cutting. It has been found that reproduceable results can be obtained if the rate of loading is within the range 0·07–0·7 MN/m^2 s. Significantly different values of uniaxial compressive strength are obtained when the above method of testing is not adhered to. In a Working Party report[3] to the Geological Society of London the following scale of strength, based on uniaxial compressive tests, was recommended:

Term	Strength (MN/m^2)
Very weak	<1.25
Weak	$1.25-5.0$
Moderately weak	$5.0-12.5$
Moderately strong	$12.5-50.0$
Strong	$50-100$
Very strong	$100-200$
Extremely strong	>200

As far as the rocks of the Coal Measures are concerned, in unconfined compression at right angles to the bedding planes sandstones tend to be stronger than siltstones and these in turn are stronger than mudstones, shales and coals (Table 2.1). The compressive strengths of sandstones with siliceous or ferruginous cement are usually greater than those with carbonate cements or clay matrix, i.e. if the amount of cement is approximately equal. When subjected to compression parallel to the bedding, a rock fails at a lower loading. Thus flaggy sandstones are weaker than massive varieties and shales weaker than mudstones. The strength of siltstones is largely dependent upon the relative amounts of quartzose and clayey material they contain, being stronger the greater the proportion of the former.

The determination of the direct tensile strength of a rock by the extension of a cylindrical specimen has proved difficult since a satisfactory method has yet to be devised for gripping the specimen without introducing bending stresses. Accordingly, most tensile tests have been carried out by indirect methods. One of the most frequently used indirect methods of testing tensile strength is the disc splitting technique which involves loading a narrow disc across its diameter. In the Brazilian test, a rock cylinder of length L and diameter d is loaded with a load P in a diametrical plane along its axis. The sample usually fails by splitting along the line of diametrical loading and the tensile strength can be obtained from:

$$T_B = \frac{2P}{\pi L d}$$

A cylindrical specimen is also used in the point-load test, being placed with its axis horizontal, between opposing cone-shaped platens and subjected to compression. This generates tensile stresses normal to the axis of loading. The tensile strength is derived by using the empirical expression:

$$T_P = \frac{0.96\,P}{d^2}$$

The length of the specimen should be at least twice the diameter of the specimen. There is a good correlation between Brazilian and point-load tensile strength and between these and uniaxial compressive strength.

When rocks are subjected to bending, shearing takes place along the bedding-plane direction. Thus the more parting planes there are in a rock, the weaker it is when so loaded. Massive sandstones and siltstones are again stronger than laminated varieties, and bending strengths of mudstones are greater than shales. The amount of bending for a given span and thickness is least in the case of sandstones. Siltstones bend less than mudstones and shales whilst these bend less than coal.

Table 2.1 STRENGTH OF COAL MEASURES ROCKS (data after Hobbs,[5] Price,[2] and Evans and Pomeroy)

Rock type	Description	Crushing strength (MN/m²)	Tensile strength§ (MN/m²)
1. Pennant Sandstone	Massive, fine-grained sandstone	168·9	18·9
2. Parkgate Rock	Massive, fine-grained sandstone	111·7	8·6
3. Tupton Rock	Massive, friable medium-grained sandstone	62·1	4·5
4. Markham Sandstone	Slightly bedded, fine-grained sandstone	108·9	10·8
5. Snowdown Siltstone	Laminated, fine-grained siltstone	93·1	6·6
6. Chislet Siltstone	Well-laminated, fine-grained siltstone	90·3	7·9
7. Seven Foot Mudstone	Slightly bedded mudstone	88·3	
8. Dunsil Mudstone	Carbonaceous, shaley mudstone	44·8	
9. Barnsley Hards Coal	—	55·7*† 34·1*‡	4·1‡ 2·8†
10. Deep Duffryn Coal	—	18·2*† 16·1*‡	0·9‡ 0·7†
11. Shale Cumbernauld	Laminated	20·4	
12. Sandy Shale Cumbernauld	Finely interbedded sandstone and shale	25·8	

* Crushed cubes (25·4 × 25·4 × 25·4 mm).
† Load applied across bedding planes.
‡ Load applied parallel to bedding planes.
§ Some of the tensile strengths in this table were derived by the disc-splitting method.

Price[4] performed a series of creep tests on the Pennant and Wolstanton Sandstones. In these tests he estimated that the maximum tensile stresses ranged from 25 to 80% of the value required to produce instantaneous failure, whereas the maximum compressive stresses were about 3 to 10% of the amount needed to produce instantaneous failure in uniaxial compression, i.e. 138 MN/m² and 252 MN/m² for the Pennant and Wolstanton Sandstones, respectively. As a consequence, he assumed that the time deflection data which he obtained from beam tests were almost totally the result of creep of rock in extension. He showed that a linear relationship existed between load and the rate of secondary creep for both these sandstones; however, the specimen of Wolstanton Sandstone which was submitted to 50% of the rupture load exhibited no secondary creep. When Price tested a nodular muddy limestone in compression (a constant load of $36 \cdot 2 \, MN/m^2$ for 70 days) he found that immediately it was loaded it underwent instantaneous elastic shortening, but the specimen quickly began to expand; this was relatively rapid in the first seven days and then slowed down. On unloading, the specimen expanded showing instantaneous and time elastic recovery. When the same sample was subjected to a second loading cycle it again exhibited instantaneous elastic shortening but this time underwent normal creep behaviour. Moreover, after unloading and complete elastic recovery the specimen was shorter than it had been immediately before the second application of loading. Price explained this discrepancy between theoretical and actual behaviour by suggesting that there had been a release of strain energy stored in the rock during and after the first period of loading and that this caused the increase in dimension of the specimen. He imagined that the retention of residual stress was not evenly distributed throughout the rock but that it was localised in pockets and he further tentatively suggested that the maximum level of stress associated with these pockets would probably be around about $69 \, MN/m^2$. Because residual strain energy stored in a specimen is not inexhaustible, when it is completely released normal creep behaviour will take place. Price also maintained that the more inhomogeneous a sample is then the more significant is the effect of residual strain energy upon the experimental time–strain results. The relative homogeneity of the Pennant and Wolstanton Sandstones, he argued, accounted for the fact that the time deflection curves obtained in the bending tests appeared to be unaffected by the release of residual strain energy. The decrease in the creep rate, noted in the experiments, was explained in terms of the progressive closing of voids and pore spaces in the specimens.

Hobbs[5] carried out a series of longitudinal strain–time measurements on cylindrical specimens of rock subjected to uniaxial compressive stresses ranging from $26 \cdot 4 \, MN/m^2$ to $41 \cdot 4 MN/m^2$ for periods ranging from a few minutes to more than a year. He found that after loading, the creep rate at first decreased and then became approximately constant. The approximate times at which primary creep ceased for three of the rocks tested were as follows:

Ormonde Siltstone	25 000 minutes
Lea Hall Sandstone	a few minutes to 5 000 minutes
Hucknall Shale	40 000 minutes

Usually the measured longitudinal time strains were smaller than the instantaneous strains which occurred when the samples were loaded. The approximate ratios of the maximum observed longitudinal creep strain to the observed instantaneous strain for the above three rocks were:

Ormonde Siltstone	1:4
Lea Hall Sandstone	1:2
Hucknall Shale	1:1

An instantaneous increase in length occurred at unloading of the specimens. This was followed by a time-dependent increase which, to all accounts, ceased after 15 000 minutes. The instantaneous increase in length was less than the instantaneous decrease which took place on loading. Hobbs accordingly assumed that both instantaneous strain and primary creep under load were not completely recoverable and that the irrecoverable strain was possibly related to applied stress.

Volumetric strain–time measurements were made on the Ormonde and Lea Hall specimens. During creep the volume of these specimens increased and their volume prior to rupture, for most of the samples which failed, was larger than the initial unloaded volume. Incidentally none of the specimens which failed under load exhibited tertiary creep prior to fracture.

Durability of Coal Measures Rocks

In a study of the disintegration of Coal Measures shales in water, Badger, Cummings and Whitmore[6] concluded that this was brought about by two main processes, namely, air breakage and the dispersion of colloid due to the dissociation of ions. It was noted that the former process only occurred in those shales which were mechanically weak whilst the latter appeared to be a general cause of disintegration. They also observed that the degree of disintegration of a shale when it is immersed in different liquids is governed by the manner in which those liquids affect air breakage and ionic dispersion forces. Therefore in a liquid with a low dielectric constant little disintegration takes place as a result of ionic forces because of the suppression of ionic dissociation from the shale colloids. It was found that the variation in disintegration of different shales in water was not usually connected with their total amount of clay colloid or the variation in the types of clay mineral present. Rather it was controlled by the type of exchangeable cations attached to the clay and on the accessibility of the latter to attack by water which, in turn, depended on the porosity of the shale. Air breakage could assist this process by presenting new surfaces of shale to water. It was also suggested that like coal, shale may also have a rank and that low-rank shales were associated with low-rank coals. The low-rank shales disintegrated most easily.

After an exhaustive investigation into the breakdown of Coal Measures rocks, Taylor and Spears[7] concluded that the disintegration of sandstones and siltstones was governed by their fracture pattern and that after a few months of weathering the resulting debris was still greater than cobble size. After that the degradation to give component grains took place at a very slow rate. The major part of their work, however, was devoted to a consideration of the breakdown of mudstones and shales. Both these types of rocks, as well as seatearths, are rapidly broken down to a gravel-sized aggregate. The mudstones and shales which were considered possessed a polygonal fracture pattern normal to the bedding. These fractures were regarded as the principal factor causing breakdown in those argillaceous rocks which were not laminated. Such fractures, together with laminations and joints, contribute towards the degradation of shales and mudstones within a matter of months. Listric surfaces in seatearths may mean that they disintegrate within a few wetting and drying cycles. Nevertheless it was noticed that a simple behaviour pattern did not exist, e.g. some rocks like the Brooch and Park seatearths, after being desiccated, disintegrated very rapidly in water (the former was literally 'explosive' and the latter broke down in less than 30 min). Although the expandable clay content in these two rocks is high and leads to intraparticle swelling, the authors maintained that this alone was not responsible for their rapid degradation. They found that breakdown could be arrested by the removal of air from the samples under vacuum. Thus they concluded that air breakage was a principal disintegration mechanism in the weaker rocks. They suggested that during dry periods evaporation from the surfaces of rock fragments gives rise to high suction pressures which result in increased shearing resistance. With extreme desiccation most of the voids are filled with air, which, on immersion in water, becomes pressurised by the capillary pressures developed in the outer pore spaces. The mineral fabric may then fail along its weakest plane exposing an increased surface area to the same process. These two authors did not dismiss the physicochemical ideas of Badger, Cummings and Whitmore[6] but suggested that such processes became progressively more important with time. They supported the concept of rank in shales.

In a later paper, Taylor and Spears[8] discussed the effect of weathering on *in-situ* Coal Measures rocks. They found that minor evidence of weathering extended to depths of approximately 6 m below the surface. On testing the highly weathered equivalents of siltstone and shale, they noted that the values of compressive strength were reduced to about one tenth of those values of the unweathered parental rocks. The values of ϕ' and c' for weathered siltstone, fragmental mudstone and seatearth, varied from 36 to 45·5°, and from 131 to 179 kN/m^2, respectively. The intergranular friction values are apparently similar to those of jointed shales and siltstones whilst the values of cohesion approach those of soft rocks. As far as the shear strength parameters are concerned, they quoted a decrease in ϕ of up to 37%, and in cohesion of approximately 93%.

REFERENCES

1. Colback, R. S. B. and Wiid, B. L., 'Influence of Moisture Content on the Compressive Strength of Rock', *Proc. Rock Mech. Symp.,* Canadian Dept. Min. and Tech. Surv., Ottawa, 65–83 (1965)
2. Price, N. J., 'The Compressive Strength of Coal Measure Rocks', *Colliery Eng.,* **37**, 283–292 (1960)
3. 'Working Party Report on the Logging of Cores for Engineering Purposes', *Quart. J. Eng. Geol.,* **3**, 1–24, Geological Society of London (1970)
4. Price, N. J., 'A Study of Time–Strain Behaviour of Coal Measure Rocks', *Int. J. Rock Mech. Min. Sci.,* **1**, 277–303 (1964)
5. Hobbs, D. W., 'Stress–Strain–Time Behaviour in a Number of Coal Measure Rocks', *Int. J. Rock Mech. Min. Sci.,* **7**, 149–170 (1970)
6. Badger, C. W., Cummings, A. D. and Whitmore, R. L., 'The Disintegration of Shale', *J. Inst. Fuel,* **29**, 417–423 (1956)
7. Taylor, R. K. and Spears, D. A., 'The Breakdown of British Coal Measures Rocks', *Int. J. Rock Mech. Min. Sci.,* **7**, 481–501 (1970)
8. Taylor, R. K. and Spears, D. A., 'The Influence of Weathering on the Composition and Engineering Properties of *In-Situ* Coal Measures Rocks', *Int. J. Rock Mech. Min. Sci.,* **9**, 729–756 (1972)

BIBLIOGRAPHY

Murchison, D. G. and Westoll, T. S., (Eds.), *Coal and Coal-bearing Strata,* Oliver and Boyd, Edinburgh (1968)
Price, N. J., 'The Influence of Geological Factors on the Strength of Coal Measure Rocks', *Geol. Mag.,* **100**, 428–447 (1963)
Raistrick, A. and Marshall, C. E., *The Nature and Origin of Coal and Coal Seams,* English Univ. Press, London (1939)
Trueman, A., (Ed.), *The Coalfields of Britain,* Arnold, London (1954)
Williamson, I. A., *Coal Mining Geology,* Oxford Univ. Press (1967)

Chapter 3

Field Investigation Techniques

In dealing with geotechnical problems in areas of mining subsidence it is axiomatic that the more that is known about the soil and rock profile affected by the mining operations the easier will be the solution of the problem. The problems themselves are dealt with in other chapters, but this chapter sets out the current field procedures in subsurface exploration in the widest sense and includes methods of exploring below the ground surface by direct, semi-direct and indirect means, the various types of sample which can be obtained and the range of *in-situ* tests which may be carried out. The results of any field investigation are processed into borehole logs and sections which indicate the general structure, the discontinuities both major and minor, and also the piezometric data. These data, combined with the results of both field and laboratory tests, are then applied to the geotechnical problems. Such problems may include the effects of mining and subsidence on existing or projected works, e.g. buildings, motorways, reservoirs and dams or underground structures, tunnels or underground water supplies.

The aim must therefore be to establish the geological geometry of the block of sub-surface crust under consideration in as detailed a manner as is necessary and to determine the parameters of the physical properties of the soils and rocks which will provide finite limits to the likely effect of the mining operations.

In any exploration programme the first step must be a search for and review of existing geological data (e.g. in the UK maps and memoirs of the Institute of Geological Sciences) as well as individual papers and mine records. In this connection a paper by Dumbleton and West[1] will be found to be of use. This should be followed by a reconnaissance survey by an engineering geologist. In such a preliminary survey any relevant exposures can be identified and mapped, and the information added to the published data. At the stage of the reconnaissance survey, existing structures should be examined and any cracking, variations from original level or line, or in fact any signs of distress, should be carefully noted and preferably photographed. From this preliminary work a general framework is built from which the detailed programme of pitting, boring, drilling, and *in-situ* and laboratory testing can be planned.

Exploration below Ground Level

DIRECT METHODS

Pits

Pitting and trenching to expose the nature and provide samples of sub-surface

40

materials is a method often neglected in the modern practice of direct investigation. There are of course limitations regarding depth. Pitting in areas where the water table is near ground level is a further limitation. Nevertheless, this method of investigation can, in certain circumstances, such as the exposure of a fault beneath thin overburden and in problems of slope stability, be a very useful method in exploration. The general limit is the depth to which modern mobile excavators can dig in one lift, which is about 5 m. If the water table is less than this depth, the effective useful depth in soils with permeabilities that allow water to rise in the pit is the depth to the water table itself. With a skilled machine operator and a competent engineering geologist to log the pit immediately after excavation and often a photographer to obtain a permanent record in colour of the pit sides and base, several 5 m pits can be excavated and backfilled in one day. The problems apart from the water table are (1) inclement weather which makes the excavation work more difficult and poses problems in logging and photography, (2) the rather scarring result to the land surface even with skilled back filling and (3) the impossibility of doing such work on very steep slopes or in very restricted spaces. Where a large number of pits are planned it is advantageous to employ two machines, one of which will excavate and the other contemporaneously backfill and clear-up the previously excavated pit. There are a number of ways of presenting the obtained information, but the pit log (Fig. 3.1) is probably the best first

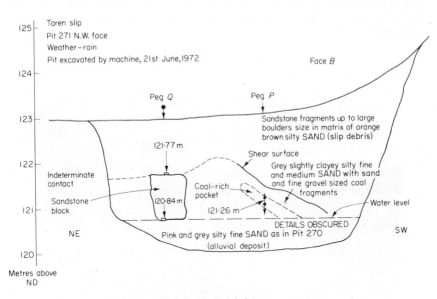

3.1. Typical trial pit log

record. With skilful organisation both disturbed and undisturbed samples can be obtained during the operations (see page 55). The question of safety should never be neglected or ignored. If the sides are likely to prove unstable, open timbering can be installed and strutted in a matter of minutes. Alternatively, where the risks are small the bucket should be placed in the base of the pit as a helpful safety precaution.

Very occasionally the problem requires a deep pit which effectively becomes a shaft with full timbering, and usually the control of ground-water either by pumping from sumps or a more elaborate system involving one or two wells outside the pit and utilising submersible pumps. Such deep excavations become minor civil engineering operations and obviously incur a very much higher cost than the simple excavations previously described. Pitting by hand is now rarely carried out except where access is restricted and where only a limited amount of work is required.

Boring

The simplest method of boring to shallow depths is to employ a hand auger which can be operated by two men. The work is hard and the ground conditions limit this method to cohesive soils without a gravel content. Sands, gravels or soils containing such materials cannot usually be penetrated. Engineering soils are therefore usually penetrated employing specially designed boring rigs (Fig. 3.2) which are often referred to as *shell and auger* rigs, though this is a misnomer carried forward from the days when the traditional boring tools were a 'shell', for use in sand and gravel, and an 'auger' for boring in clay. Most investigations are now carried out using 'drop tools' attached to a wire rope or cable and a more correct description is *cable tool boring* rig, the two basic tools being a 'shell' and a 'clay-cutter'.

3.2. Investigator boring rig being employed in conjunction with a hydraulically operated pendant rotary drill

The winch capacities of these rigs usually vary between 1 and 2 t. The maximum depth of borehole depends on the nature of the ground being investigated and it is difficult to classify rigs in terms of depths of hole. Using several reducing sizes of casing boreholes of 60 m depth are not unusual.

The basic tools (the 'shell' and the 'clay-cutter') are open-ended steel tubes of varying lengths and diameters with specially designed separate cutting shoes. By fitting the appropriate cutting shoe the tool can be used as a shell or clay-cutter.

For boring in stiff clays and highly weathered rocks the weight of the tool can be increased by adding an extra section of steel tube or by the addition of 'sinker bars'. The use of excessively heavy tools in soft clay soils should not be permitted as this is likely to cause disturbance below the base of the borehole.

To penetrate weak rocks, boulders and large cobbles, heavy chisels are used. They are smaller versions of tools used for water well boring and can weigh up to 250 kg. Progress is slow as the material has to be removed with a shell after it has been broken-up by the chisel.

When boring in soft clays the borehole is unlikely to remain open without the support of steel lining tubes or casing. Although the hole may not collapse completely, the tendency is for the sides to squeeze inwards and jam the shell. The usual practice in such clays is for the operator to bore ahead of the steel casing for a distance of 1·5 m (the standard length of a casing section) before adding a new section of casing and surging it down. The reason for surging the casing is to keep it 'free' in the borehole so that it can be easily extracted on completion. When the operator can no longer advance the casing by surging he will reduce to a smaller diameter with the advantage of having 10–15 m of the smaller size casing free-standing inside the larger hole. This reduces the frictional resistance of the casing and the boring progress generally increases. The decision to reduce to a smaller size is not taken lightly. The operator has to decide whether the reduction in boring time justifies the delays incurred in running-in the extra set of casing and extracting it on completion. If the hole is near its final depth he may resort to 'driving' the casing with the clay-cutter and 'jacking-out' the casing on completion.

In stiff clays the borehole can very often be advanced without lining tubes, and within the time required to make the boring the borehole often remains dry. A short length of casing is used at the top of the boring to keep the hole stable and prevent collapse. Where clays occur below granular deposits the casing used as a support in the granular soils is driven a short distance in the clay to create a seal and the shell is used to remove any water which might enter the borehole. In very stiff clays a little water is often added to assist boring progress. This must be done with caution to avoid possible changes in the properties of the soil to be sampled.

With very few exceptions sands and gravels always require the use of lining tubes to support the sides of the borehole. The casing must be advanced with the borehole or the continual collapse of the sides of the borehole below casing will prevent further progress and result in a cavity being formed (a condition known as *overshelling*). Cases have been known where overshelling at depths of up to 10 m has caused a depression to occur at ground level. The piston action of the shell causes a loosening of the material at the base of the borehole and the casing is driven into this disturbed zone. Because of the mode of operation of the shell the borehole must be full of water for the shell to operate efficiently. Since most granular strata in the UK are water bearing all that is required is a supply of water to keep the natural water level in the borehole 'topped up'. The flow of water must always be from the borehole into the surrounding strata. If this condition is reversed by allowing water to flow into the borehole, 'piping' will probably occur, and will invalidate the results of standard penetration tests (page 56). *Piping* is the term used to describe the condition where material is carried up the borehole when the hydrostatic head in the borehole is less than the hydrostatic head in the surrounding ground. Provided the head of water in the borehole is greater than the natural head, piping can usually be prevented. To overcome artesian heads the lining tubes must be extended above ground level and kept filled with water.

Coarse gravels above the natural water table present special problems. Because of their high permeability it is usually impossible to maintain a head of water in the borehole. Consequently, the shell cannot be used. Fortunately these conditions usually occur at or near ground level and the problem can sometimes be overcome by using an excavator to open a pit either to the water table or to a depth of 3–4 m. Lining tubes can then be inserted, the pit backfilled and boring can then proceed normally through the lining tube. Another method of penetrating gravels and cobbles above the water table is to employ a special grab with a heavy tripod and winch, and employ casing of 400 mm diameter or greater.

The importance of ground-water and piezometric data and permeability is very often under-recognised in exploration of soils and rocks. This question will be dealt with more fully later in the chapter, but it is important that the boring operator should receive clear instructions which must be rigorously carried out. The basic instructions should be (1) record the depth at which water was first noticed in the borehole, (2) if water is added to the borehole to assist the boring operation it must be noted, (3) the level of the water and the amount and diameter of casing in the boring at the end of the shift must be noted and (4) the level before work commences on the following morning must also be noted. Requirements 3 and 4 are more important than requirements 1 and 2. Failure of the boring operator to observe these instructions limits the value of the boring.

Rotary Drilling

The only direct method of penetrating rocks, other than weak and very weathered rocks which can sometimes be more advantageously treated as soils and explored by the boring techniques already described, is to employ a rotary drill. In the last two or three years hydraulic rotary attachments have been developed that can be operated in conjunction with a percussion boring rig.

A typical medium rotary drill is shown in Fig. 3.3. A comparison of the maximum torque for both types of drills shows that the rotary attachments have only about one third of the power of the self-contained drills. This seriously limits their drilling capacity and they are normally only used for short core runs to prove rock at the base of a borehole.

Standard machines are available as skid-mounted or trailer-mounted units. Some larger units are often lorry mounted. A variety of core barrels are available for rock sampling, but in their simplest form they consist of a steel tube with a cutting bit at the lower end. This bit is set with diamonds or tungsten carbide 'inserts' and as the barrel is rotated the bit cuts (or grinds away) an annulus of rock allowing the core to pass up into the barrel. A flushing fluid, generally water but increasingly air and sometimes mud, is pumped through the drill rods and discharged at the lower end of the core barrel. The purpose of this is two-fold, firstly to act as a coolant for the bit and secondly to remove the cuttings from the drill hole. A cross-section of a typical double tube core barrel is shown in Fig. 3.4. Good core recovery, which must be the aim of every contractor, is dependent on the selection and use of the correct type of core barrel and core bit.

The main types of core barrels are as follows:

3.3. Medium rotary drill

Single Tube   This is the simplest type of core barrel and suitable only for massive uniformly hard rocks. It is rarely used in practice.

Double-Tube Rigid   This type of core barrel has an outer and inner tube both of which are fixed to the core barrel head. A reamer shell and bit are screwed on to the outer tube. The flushing water passes between the two tubes, through holes at the bottom end of the inner tube, then alongside the core before passing around the face of the bit. It is only suitable for hard formations and has the advantage that this is the type least likely to jam.

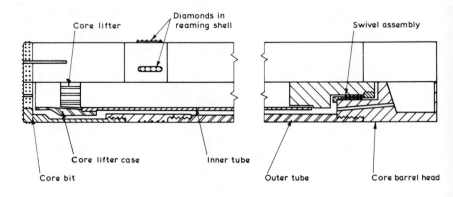

3.4. Double-tube swivel-type core barrel

Double-Tube Swivel   One disadvantage of the double-tube rigid barrel is that both the inner and outer tubes rotate together and in soft rock this can break the core as it enters the inner tube. The double-tube swivel-core barrel incorporates a bearing in the core-barrel head, which permits the outer tube to rotate while the inner tube remains stationary. In other respects it is similar to the double rigid type. It is suitable for medium and hard rocks and the non-rotation of inner tube does give improved core recovery even in soft friable rock formation.

Face-Ejection Double-Tube Swivel   Although the double-tube swivel type core barrel has advantages over the rigid type, it has the disadvantage that for a few inches below the end of the inner tube the core is washed by the flushing fluid. In the face-ejection type, an extension, known as the core *lifter case*, is fitted to the bottom of the inner tube, extending almost to the diamond bit. The flushing fluid passes down between the two tubes and out through ports in the face of the bit, eliminating the eroding effect of the water on the core. The barrel is also known as a *face-discharge* core barrel. The face-ejection barrel is a minimum requirement for coring badly shattered, weathered and soft rock formations.

Triple Tube   A newer development is to employ a core barrel with an inner triple tube which is split into two halves longitudinally. Thus when withdrawn from the outer casings of the core barrel the core can be observed and described without the inevitable risk of disturbance occasioned by hydraulic extrusion. This procedure permits a much better evaluation of highly cleaved and jointed rocks and should become standard in the years to come in high quality investigations.

Retractable Triple Tube   This has a conventional outer tube with the inner tube spring mounted. This barrel was specially developed to core very soft rocks, especially those containing or alternating with stronger rock formations. When drilling through soft rock the inner barrel is extended beyond the face of the standard bit. This allows the core to be cut dry as it is completely protected by the extended inner barrel. When harder material is encountered the increased pressure causes the inner barrel to retract and the coring continues in the conventional manner.

Disturbance of the core is most likely to occur when it is being removed from the core barrel. Most rock cores should only be removed whilst the inner tube is held with its long axis in a horizontal plane. Specially designed hydraulic extruders can then be used to remove the core. To reduce disturbance during extrusion the inner tube of double core barrels can be lined with a plastic sleeve before drilling commences. On completion of the core run the plastic sleeve containing the core can be withdrawn from the barrel enabling the geologist to examine the core in as natural a condition as possible. The widespread practice of using a hammer to tap out the core from the inner barrel direct into the core box is not satisfactory and should be discouraged.

Most core drilling is carried out using diamond bits which are available in two main forms: (*a*) 'surface set' bits with individual diamonds set in a metal matrix and (*b*) 'impregnated bits' with fine diamond dust incorporated in a matrix. The diamonds used for the surface set bits vary in both quality and size. Choice is governed by the rock to be drilled. Basically it can be summarised as 'the harder the rock the smaller the size and the higher the quality of the diamonds'. Surface set bits should be taken out of service when about 25% of the diamonds are lost or broken. Beyond this point the cutting rate decreases rapidly. The remaining diamonds are leached out of the matrix for resetting in new bits. Impregnated bits have no salvage value. A 'bit selection chart' is given in Table 3.1. The chart is intended as a broad classification only. Bit performance (and consequently drilling progress) depends on several factors including the peripheral speed of the bit, the applied pressure, the hardness of the matrix and the shape and quality of the diamonds.

Tungsten bits are less costly than diamond bits but generally are not suitable for drilling in very hard rocks.

The dimensions of core barrels and bits are given in Table 3.2.

Mention should be made of the large rotary drills which at low revolutions can operate continuous flight augers in soils. These are finding an increasing function where large diameter cores *UF* and *ZF* (140 and 165 mm) are required in rocks, though their use as augers has its limitations.

In considering these large drills when operating augers, the action of screwing hollow-stemmed augers into the ground rather than driving casing was thought to have certain advantages, notably a rapid boring rate and the ability to obtain undisturbed samples ahead of the augers through the hollow stem. In certain ground conditions, particularly granular materials below the water table, a number of operating difficulties have been encountered and these still remain unsolved. Due to this and the possibility that without continuous sampling there is a risk that critical changes in ground conditions may be missed, power augers are limited in their use in site investigations.

Augers are available as 'solid-stem' or 'hollow-stem', both having an external continuous helical flight. The hollow-stem augers are used in soils where the borehole will not remain open without the aid of a steel lining. The hollow central stem is sealed at the lower end with a special combined plug and cutting bit which can be removed when a sample has to be taken. To allow clearance of the sampling equipment the internal diameter should not be less than 150 mm. The solid-stem augers, usually of 150 mm o.d. are used in stiffer clays where casing is not required. Solid augers have to be withdrawn from the borehole each time an undisturbed sample is required.

Table 3.1 RELATIONSHIP OF ROCK TYPES AND SIZE OF DIAMOND IN DIAMOND BITS (bit selection table)

Category	Typical rock types	Diamond sizes (*stones per carat*)
I	Chalk Marl Shale	2–8 stones/carat or sawtooth, or tungsten carbide
II	Sandstone Limestone Siltstone Graywacke Slate Dolomite	Generally 8–16 stones/carat or sawtooth bits, but sometimes 16–30 stones/carat are preferable
III	Schist Gneiss Some Sandstones and Limestones	30–60 stones/carat or impregnated bits
IV	Granite Basalt Quartzite Rhyolite and similar igneous rocks	60–100 stones/carat or impregnated bits

Disturbed samples taken from auger holes are often unreliable. If a sample is taken from the base of the auger after it has been removed from the borehole it is likely that it is contaminated by material from the sides of the hole. If the auger is held stationary at a particular level and rotated rapidly it will act as an 'Archimedean screw' and tend to force material up the flights to the surface. Again this material is likely to be contaminated and there is no guarantee that it has come from a particular level.

Hollow-stem augers are useful for investigations where the requirement is to locate and prove bedrock. The hollow-stem augers penetrate the overburden to bedrock. The plug at the base of the auger is then withdrawn to allow conventional rotary drilling equipment to be used to core the rock.

In favourable ground conditions, such as firm and stiff homogeneous clays, auger rigs are capable of high output rates. A high degree of technical supervision is required to ensure that the quality of sampling is unaffected by the high production rates. Normally a senior soils technician is required to provide full time supervision for each unit which in addition requires a somewhat larger labour force than that required for a 'cable tool' rig. The capital cost compared with other boring and

Table 3.2 GENERAL DIMENSIONS OF CORE BARRELS, CASING AND RODS*

Core barrels				Drillcasing			Drill rod		
Design		Core dia. (mm)	Hole dia. (mm)	Designation	o.d. (mm)	i.d. (mm)	Designation	o.d. (mm)	i.d. (mm)
AWX	AWM	30·2	49·2	AX	57·2	48·4	AW	44·5	15·9
BWX	BWM	41·3	60·3	BX	73·0	60·3	BW	54·0	19·0
NWX	NWM	54·0	76·2	NX	88·9	76·2	NW	66·7	34·9
HWX	HWF	76·2	100·0	HX	114·3	100·0	HW	88·9	60·3
—	PF	92·1	120·6	PX	139·7	123·8			
—	SF	112·7	146·0	SX	168·3	149·2			
—	UF	139·7	174·6	UX	193·7	176·2			
—	ZF	165·1	200·0	ZX	219·1	201·6			

* After BS 4019, Pt. 1, 1966 (under revision).

drilling rigs is high. However, as already stated, one very useful function of these heavy rigs with their high torques is to employ them as rotary rigs for obtaining large diameter cores. These large diameter cores ($ZF = 165$ mm) are usually necessary where heavy foundation loads are envisaged in a weak rock environment or where a more detailed appreciation of the rock fabric is required.

Auger rigs are usually tractor mounted for mobility although some earlier models were mounted on a lorry chassis. Experience has shown that in the British climate lorry-mounted units become bogged down very easily and delay factors of up to 30% of operating time are not uncommon.

The logging of boreholes and drillholes should be carried out with the minimum delay and ideally an engineer or engineering geologist should describe the samples at the site as they are removed from the borehole or drillhole. This is a practice which is becoming more widespread as, apart from improving the quality of description, the boring operator can receive instructions to vary his technique, sampling type and interval as the borehole progresses and a much higher quality of data results. However, where this is not possible all samples and cores should be transported with care to a laboratory or sample description building and described there. From the sample/core description sheets, the logs are produced and generally the form should be as shown in the example given in Fig. 3.5. With regard to the logging of rock cores in particular, the procedure recommended by the working party of the Engineering Group of the Geological Society should be employed[2]. However, as this involves a large amount of detail a very recent practice is the production of additional summary logs which permit an earlier appreciation of the fundamental points and avoid the difficulty of not being able 'to see the wood for the trees'. The detailed logs are retained for closer study where this is necessary. A simple section made up from three boreholes on a section of motorway in a mining area of south Wales is shown in Fig. 3.6. A word about numbering; an unique number must be used for each borehole, drillhole or pit. Put another way, the same number or even the same number with a prefix or suffix should preferably not be used on the same site, otherwise doubts and confusion are likely to occur. A good principle is to employ the 100 series for pits, the 200 series for boring or drilling and the 300 series for soundings.

SEMI-DIRECT METHODS

The technique of driving into the ground a rod with or without a specially dimensioned toe or cone and recording the driving resistance has long been employed as a semi-direct means of exploring sub-surface conditions. The simplest technique has been developed as the Mackintosh tool in which light rods can be driven-in by two men employing a hammer or 'monkey' for driving and withdrawing the rods. It has its application in exploring materials such as peat, or to give a rough appreciation of the depth to rock where the overburden is thin.

The Dutch, however, over the last 30 years or so have developed advanced sounding techniques which are particularly suited to the Dutch geological conditions of fine-grained soils (silts and sands) with bands of gravel as pile-bearing strata. Such conditions occasionally occur in the UK, notably in estuarine areas. In this technique a tube and inner rod are advanced into the ground hydraulically, the reaction being obtained from screwed-in pickets. The rod has a specially designed steel cone at the base with a cone angle of 60° and a cross-sectional area of

Borehole No 212

Sheet 1 of 2

EQUIPMENT AND METHODS	Location No. 6244
Cable tool boring 200mm dia. to 6·8m, then water flush, 60mm diamond rotary core drilling to 8·3m	SOUTH PEMBROKESHIRE SCHEME

Carried out for	Ground level	Co-ordinates	Date
L & D Consulting Engineers	+11·32	3817 4392	30/8/73 to 31/8/73

Description	Reduced level	Legend	Depth & thickness	Samples/tests Depth	Sample Type	No.	Test	Field records
Topsoil	11·32		(0·2)					
	11·12		0·20					
Loose brown medium sand with some fine sand and occasional sub-angular fine to medium gravel				1·00-1·45 1·70	D	1	S	N= 6
					D	2		
			(4·30)	2·00-2·45	D	3	S	N= 7
				2·70	D	4		
				3·30-3·45	D	5	S	N= 8
				3·70	D	6		
	6·82		4·50	4·00-4·45	D	7	S	N= 8
Soft grey clayey silt with angular sandstone cobbles towards base			(1·70)	5·00-5·45	U	8		24 blows for U100 penetration
	5·12		6·20	6·15	D	9		
Weathered red-brown very weak fine grained micaceous sandstone	4·52		(0·60) 6·80				C r 75 0	Good water returns throughout

Main description	Details							
Thickly bedded fresh brown-red strong fine grained micaceous Sandstone	6·80-7·80 Shattered with some red sandy silty clay 7·80-8·30. Frequent fractures at 20°–40° to horizontal		(1·50)	Penetrated			60 0 65 20	
	3·02		8·30					
End of borehole								

SPT	Where full 0.3m penetration has not been achieved the number of blows for the quoted penetration is given (not N – value)
Depths:	All depths and reduced levels in metres. Thicknesses given in brackets in depth column.
Water:	Water level observations during boring are given on last sheet of log

Sample/Test Key
D Disturbed Sample
B Bulk Sample
W Water Sample
I Piston (P),Tube (U) or core sample; Length to scale
S Standard Penetration Test (N)
V Vane Test
C Core recovery (%)
r Rock Quality Designation (RQD–%)

Remarks
1* Core recovery assessed from distributed fragments
2. Water was added to assist boring to 6·8m depth

Logged by J.A.F.

Scale 1:50

Fig. 24

3.5. Borehole log

1000 mm². At about 300 mm intervals the cone is advanced ahead of the tube for 50 mm and the maximum resistance on a load cell is recorded. The tube is then advanced and 'catches up' with the cone and the two are advanced further and the process repeated. A modified cone is available that enables the frictional resistance of a short length of tube immediately behind the cone to be measured. Machines of 2, 10 and 17 t capacity are available. In Fig. 3.7 the 17 t reaction sounding rig is shown operating the Delft continuous sampler (q.v.). A more recent development has been the introduction of an electrical recording device which produces a continuous plot of penetration resistance as the test proceeds. Other applications of the test are to prove the thickness of soft deposits overlying more compact strata or to confirm the soil profile between boreholes. A small 25 mm sampler (the 'Shrew') can be used in conjunction with the sounding machine to take very small samples for identification purposes.

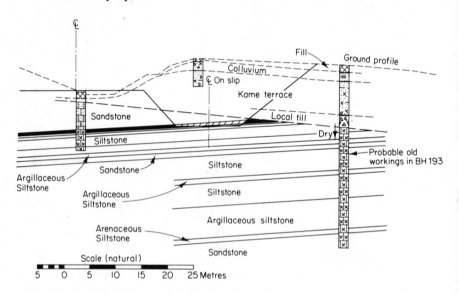

3.6. Section of a motorway investigation in a mining area of south Wales

A modification of the static or 'Dutch' sounding technique is dynamic/static sounding, whereby the cone is driven dynamically between test levels. The method is often used when the load required to overcome the frictional resistance of the drive tubes is greater than the holding power of the screw pickets used to provide the reaction. It enables this essentially Dutch technique to be employed where, for example, relatively thick gravel deposits which cannot be penetrated by static thrust alone occur at a high level overlying sands and silts. A typical sounding log is given in Fig. 3.8.

The advantages of the soundings are (1) that they can give a rapid appreciation of ultimate loads for end bearing piles, and (2) that they give a rapid correlation of ground conditions with borings. However, the sampling possible is very limited and restricted to small disturbed samples. The rigs have a dual function in that they are necessary in the employment of a specialist continuous-sampling technique in alluvial fine-grained soils (Delft sampling, q.v.).

3.7. Sounding rig (17 t reaction) employed with the Delft continuous sampler

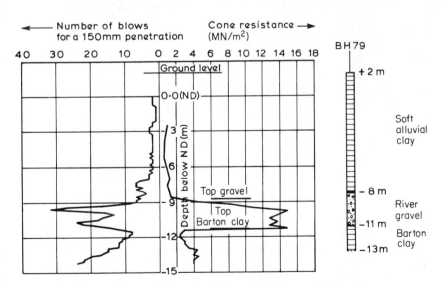

3.8. Sounding log static/dynamic technique at Fawley, near Southampton

INDIRECT METHODS—GEOPHYSICS

Although magnetic and gravitational methods can be employed in the type of exploration being considered, the usual methods which have the greatest application are the resistivity and seismic techniques, which are dealt with in Chapter 4.

Ground-water

It is astonishing how little attention is paid to the determination of ground-water conditions in exploration. All too frequently in examining reports it is apparent that the piezometric data are too meagre and even that which has been obtained is probably of doubtful value. It is obvious that in homogeneous relatively permeable ground there will be an easily determinable water table with the hydrostatic head increasing at a linear rate below this level. However, structure and varying permeabilities lead to variable water pressure conditions at different depths. A standpipe consisting of a perforated tube placed in a borehole and surrounded by gravel to near the surface, with a seal of 1 m to keep out surface drainage, will record only the upper free water surface. For determining the water pressure within individual strata or bands, piezometers must be sealed into the zone from which such information is required. A diagram of typical piezometers is shown in Fig. 3.9. The installation is a matter for skilled technicians, and attempts to install

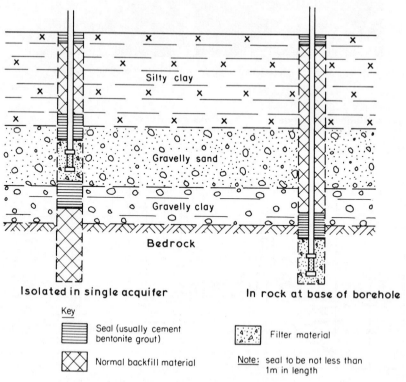

Isolated in single acquifer   In rock at base of borehole

Key

Seal (usually cement bentonite grout)

Normal backfill material

Filter material

Note: seal to be not less than 1m in length

3.9. Typical simple piezometer installations

more than one piezometer in any one boring increases the failure rate (to function properly) from less than about 5% to a much higher figure. The number of piezometers required will vary with the problem, but generally a minimum of half the borings and drillings carried out should be converted to observation wells or piezometers. Once installed their ability to function should be checked by topping up and then they should be observed daily for a period until an apparent equilibrium has been reached. Thereafter the intervals between observations can be lengthened to a week or a month but ideally should continue for about a year.

Methods of Sampling in Soils

DISTURBED SAMPLES

Representative disturbed samples taken from the clay cutter or shell should be sealed in glass or plastic jars of a sufficient size to permit index testing, i.e. about 0·5 kg in weight. With coarse granular materials, if the particle size distribution is required then obviously a larger sample is necessary and this may be retained in a tough plastic sack. With such samples of granular soils care must be exercised to avoid loss of 'fines' when taking samples. The whole contents of a shell (repeated two or three times) should be emptied into a suitable container and allowed to settle before pouring off the water then mixing the material by hand before taking a representative sample.

UNDISTURBED SAMPLES

The standard undisturbed sample taken in cohesive soils is the driven tube approximately 100 mm in diameter and therefore designated *U100*. It is about 450 mm in length and the typical details are given in Fig. 3.10. The cutting shoe should conform to the limits given in British Standards Code of Practice CP 2001 *Site Investigations*[3].

This type of sampler is ideal for clays with a shear strength greater than 50 kN/m^2. For soils of lower shear strengths, thin-walled piston samplers are preferable. These are available in sizes ranging from 54 to 250 mm in diameter. Piston samples are taken by advancing the sampler by means of hydraulic or

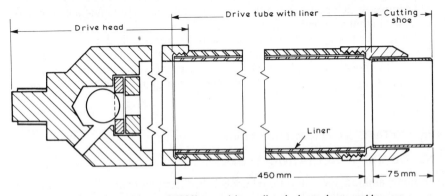

3.10. 100 mm diameter (*U100*) open-drive undisturbed sampler assembly

mechanical jacks or by a pulley system. The piston which seals the bottom end of the sample tube is locked into place until a sample is to be taken. This allows the sampler to be advanced beyond the base of the borehole, into undisturbed material, before the piston is released. Dynamic driving of piston samplers should not be carried out.

A very important factor likely to affect the quality of samples is accurate control of sampling depths. Before an undisturbed sample is taken, or an *in-situ* test carried out, the depth to the base of the borehole must be checked by careful measurement. Any initial penetration of the sampler under its own weight, or the weight of the drill rods, must be taken into account to avoid overdriving which will lead to compaction of the sample and produce unreliable test results.

In soft sensitive clays and silts the use of shells to advance the borehole may result in disturbance of the soil at the base of the borehole. Whenever possible the last 500 mm of soil above the sampling position should be removed by augering. If this is not possible boring should be stopped about 500 mm above the sampling position. The piston sampler can then be advanced beyond the base of the borehole, to the sampling position, before the piston is released. In very soft soils continuous piston samples can be taken without advancing the borehole. When the frictional resistance prevents the sampler being advanced any further the borehole can be cleaned out to a point about 1 m above the last sample position and further sampling carried out. This method greatly reduces the risk of sample disturbance. Disturbance can also be caused if the water level in the borehole falls below the natural ground-water level. To avoid possible disturbance the water level in the borehole must be maintained above the ground-water level.

Both *U100* and piston samples should be sealed with paraffin wax in a just-molten state and the sample then capped. The transport to a laboratory, particularly of piston samples, should be carried out in such a way that disturbance is minimal. Storage should be permitted only in frost-free conditions.

For investigations involving significant thicknesses of alluvial soil where details of the soil fabric assume a greater importance than usual, a continuous sample of 29 or 88 mm diameter can be obtained from ground level to depths of about 20 m by employing the sounding units already mentioned for advancing a sampling device into the soil which cuts and retains a continuous core of soil within a self-vulcanising sleeve as the sampler is pushed further into the material being sampled. This Delft sampler (so called because it was developed and perfected by the Laboratorium voor Grondmechanica at Delft, Holland) postdates the Swedish foil sampler which performed much the same function in retaining a continuous core, but in that case within aluminium foil.

Undisturbed samples of non-cohesive soils up to coarse sand size (600μm) can be taken with the 'compressed-air sand sampler' which incorporates a 60 mm diameter thin-walled sampling tube. As the procedure is slow and the cost of such samples is high, the method is therefore restricted to investigations where such samples are a vital requirement.

In-situ Testing and Instrumentation

STANDARD PENETRATION TEST

In sands and gravels conventional undisturbed samples cannot be used and the problems are usually dealt with by considering results of *in-situ* standard penetration tests (s.p.t.s) with index testing of representative bulk samples. Bulk samples are collected from the materials recovered from the 'shell', care being taken to avoid the loss of fine material. Reliable representative samples of sands and fine gravels can often be obtained by using a standard-penetration-test tool with an open shoe. These samples cannot be regarded as undisturbed because of the thickness of the cutting shoe, but they provide a reliable indication of the particle size.

The standard penetration test is an empirical test used in sands and fine gravels to determine in the first instance the relative density of the materials being penetrated. The relative density refers to the degree of compaction of the material, i.e. loose, medium dense or dense, and must not be confused with the bulk density. The test requires that a 50 mm o.d. sampler (Fig. 3.11) is driven into the base of the borehole using a standard weight (63·0 kg) falling *freely* through a distance of 750 mm. The sampler is initially driven 150 mm to ensure that it has passed through the material disturbed by the shell. The number of blows required to drive it a further 300 mm are recorded. This blow count, known as the N value, can be related to a series of design curves to determine the size of foundation required to support a given load within acceptable limits of settlement. The position of the water table, the depth of the foundation and the grain size of the material tested, must be taken into account in analysing the results of the tests. For further information related to the interpretation of the results of standard penetration tests, reference should be made to Tomlinson[4].

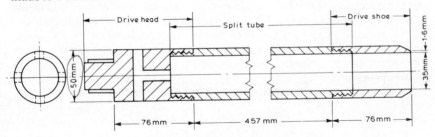

3.11. Standard penetration sampler assembly

Since the penetration test is an empirical test it is essential that every effort is made to carry it out in as standard a manner as possible. To achieve this the weight used to drive the sampler is now incorporated in a semi-automatic trip-hammer which ensures a free-fall drop of constant height. Caution must be exercised to avoid carrying out the test in disturbed or piped material. If there is any doubt or if the blow count is less than 10, the sampler may be driven an additional 300 mm in order to permit a more realistic assessment of the N value. In certain investigations of granular soils it is advantageous to extend the information obtained from standard penetration tests by carrying out cone sounding.

The standard penetration test can also be employed in clays, in weak rocks and in weathered zones of harder rocks. The interpretation of the results is a matter of experience and correlation with other data.

STATIC AND STATIC/DYNAMIC SOUNDING

The method of operation of this *in-situ* testing technique has been described under

Semi-direct Methods (page 50). The cone resistance in granular soils gives a direct reading of the ultimate resistance for end bearing piles. The use of soundings as a rapid method of correlation between boreholes is particularly useful. The results in granular soils for assessing the relative density and ϕ, the angle of shearing resistance, are a further benefit. In soft clays the relationship between the undrained shear strength and the cone resistance requires correlation by measurement of the sheear strength from laboratory or *in-situ* vane shear tests. In stiff clays, penetration for more than a very short distance is not possible. The value of this technique is therefore at its best when employed in silts and sand where the sands tend to be loose and variable, i.e. in estuarine conditions or in areas of recent alluvium. A very full appreciation of sounding and other penetrometer techniques, and interpretation, has been given by Sanglerat[5].

VANE TESTS

Although soft sensitive clays can be sampled with standard piston samplers, possible disturbance between the site and testing laboratory can often cause doubts regarding the validity of results of triaxial tests carried out on such samples. The vane test is therefore often used to measure *in-situ* the undrained shear strength of clays within the consistency range of very soft to firm.

In its simplest form the vane apparatus consists of four blades arranged in a cruciform and attached to the end of a rod. The vane and rods are pushed into the soil and the assembly rotated. When the vane is rotated the soil fails along a cylindrical surface swept out by the edges of the vane as well as along the horizontal surfaces at the top and bottom of the blades. If the dimensions of the vane are known the shear strength, at failure, can be determined from the measured maximum torque applied to the vane rod.

For a vane of height H and diameter D it can be shown that:

$$M = \frac{\pi}{2} HD^2 S_V + \frac{\pi}{6} D^3 S_H$$

where M = maximum torque,

  S_V = shear strength in vertical direction, and

  S_H = shear strength in horizontal direction.

If $H = 2D$, which is commonly the case, then

$$M = \pi D^3 (S_V + S_H/6)$$

and if it is assumed that the shear strength is the same in all directions, i.e. $S_V = S_H = S$, then

$$M = \frac{7\pi D^3 S}{6}$$

$\therefore$ $M = 3 \cdot 665 \, D^3 S$ (3.1)

In modern vane equipment (Fig. 3.12) in order to eliminate the friction of the soil on the vane rods during the test, all rotating parts, other than the vane, are enclosed in guide tubes. The vane rods, which pass through guide tubes, are supported on a thrust bearing to minimise, as far as possible, the friction induced by their self weight. The vane is normally housed in a protection shoe which incorporates seals to prevent the ingress of soil particles into the thrust bearing.

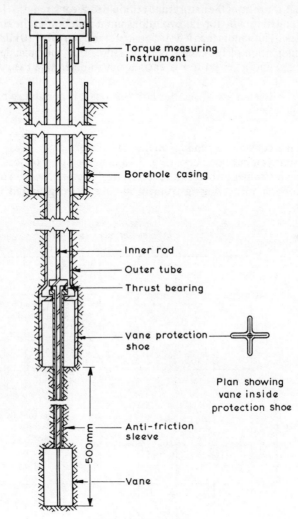

Torque measuring instrument

Borehole casing

Inner rod
Outer tube
Thrust bearing

Vane protection shoe

Plan showing vane inside protection shoe

500mm

Anti-friction sleeve

Vane

3.12. Vane test apparatus

For each test the vane assembly is advanced into the soil, either from ground level or from the base of the borehole, to a point 0·5 m above the test position. The vane is then pushed out of the protection shoe and advanced to the test position. The torque is applied to the vane rods by means of a torque-measuring instrument mounted at ground level and clamped to the borehole casing or rigidly fixed to the ground. Ideally the mechanism should be rotated by a worm and pinion gear driven by a hand wheel or crank. The recording device, which normally measures the displacement of a spring, should incorporate a maximum-reading indicator and allow for adjustment of the zero setting.

During the test the vane is rotated (with a rate of strain of 6–12°/min) until the soil is sheared and the maximum reading is recorded. The natural undrained shear strength can then be calculated. It is often desirable to determine the sensitivity of the soil, i.e. the ratio of the natural and remoulded shear strengths. In such cases the vane is rotated rapidly for 12 revolutions immediately after the completion of the natural test. The vane is then left stationary for an interval of 10 min to allow for the partial dissipation of the excess pore pressures that will have developed during remoulding, and a repeat test is carried out to determine the remoulded shear strength.

Since most vanes are of standard dimensions it is possible to rewrite equation 3.1 as follows:

$$M = KS$$

where K is a constant depending on the vane dimensions.

A standard calibration curve (Fig. 3.13) can then be produced for the torque-measuring instrument whereby the recording device can be read in units of shear strength. Each measuring instrument should be recalibrated at six-monthly intervals.

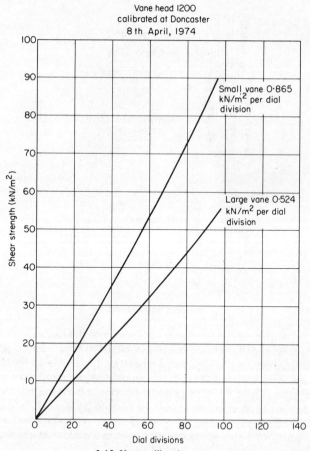

3.13. Vane calibration curve

Tests in clays with a high content of organic material, or pockets of silt or sand, are likely to produce erratic high results. It follows, therefore, that vane tests should not be regarded as a means of investigation in isolation but should be carried out in conjunction with boreholes and interpreted by experienced engineers. Further details on vane testing and its application will be found in References 6 to 8.

PERMEABILITY

The rate at which ground-water under pressure will flow through the soil is of importance in a number of geotechnical problems. Mass permeability is best determined by *in-situ* measurement.

A preliminary assessment of the magnitude and variability of the coefficient of permeability can be obtained from tests carried out in boreholes as boring or drilling proceeds. By artificially raising the level of water in the borehole above that in the surrounding ground the rate of flow of water from the borehole into the surrounding ground (falling-head test) is measured. Conversely, the water level in the borehole can be artificially depressed so that the rate of flow of water into the borehole (rising-head test) can be determined. In the falling-head test it is likely that the fine particles suspended in the water may settle at the bottom of the hole which will reduce the measured permeability in relation to the true ground permeability. In the rising-head test there is a risk that the upward flow of water in sand and fine gravels will loosen the material to some depth immediately below the base of the hole, which will result in a measured permeability which will be higher than that of the undisturbed ground. Additionally, in cohesive soils the results can be affected by the smearing action of the boring tools.

Even with these limitations *in-situ* tests of this type in boreholes have useful applications. However, for the results to be of value the observations have to be made, recorded and interpreted with particular care.

If sufficient time is allowed to elapse for the steady state to be approached the permeability k can be evaluated by measuring the rate of flow of water q under a constant applied change in head H, using the following simple expression given by Hvorslev[9]:

$$k = \frac{q}{FH}$$

where F denotes the shape or intake factor. For the simple case of a lined borehole of constant diameter D within a uniformly permeable stratum, the shape factor F is $2 \cdot 75\, D$. The shape factors for other practical situations are given by Hvorslev[9].

For the more general case of a rising- or falling-head test in which the piezometric head H within a borehole will vary with time t as water enters or leaves the hole, assuming, a constant ground-water level, the permeability k for variable head tests is determined from the expression

$$k = \frac{A}{F(t_2 - t_1)} \times \log_e \left(\frac{H_1}{H_2} \right)$$

where H_1 and H_2 are the piezometric heads causing flow at times t_1 and t_2 respectively, and A is the inside cross-sectional area of the borehole casing.

The test procedure observes the level of the water within the casing at measured elapsed times from the commencement of the test, and the results should be tabulated (Table 3.3). The results are subsequently plotted on the graph of water level (plotted as ordinate) against elapsed time (abscissa) to check the consistency of the results, and this being so a smooth curve would be drawn through the points (Fig. 3.14).

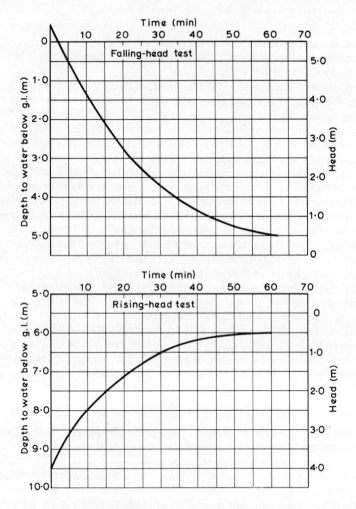

3.14. Rising- and falling-head permeability tests

In cohesive soils of low permeability it is preferable to carry out the test over an unlined section of borehole so that a more rapid change of water level can be achieved during the test period. In firm to stiff clays this should not present any

Table 3.3 FIELD PERMEABILITY TEST

Loc. No.:—
Borehole No.:—
Date of test:—
Depth of borehole before test: 11·0 m
Depth of borehole after test: 11·0 m
Depth of casing below ground level: 11·0 m
Height of casing above ground level (H_1): 0·50 m
Diameter of casing: 200 mm
Diameter of boring tool: 175 mm
Depth of ground-water (if known) (H_4): 5·50 m

Name:—
Reduced level:—
Carried out by:—

Elapsed time (min s)	Depth to water from top of casing (H_2) (m)	Depth to water from ground level (H_3) ($H_3 = H_2 - H_1$) (m)	Head H (see Note 3) (m)	$\frac{H}{H_0}$ (see Note 4)
0 0 = t_0	0·0	10·00	4·50 = H_0	1·0
0 30	10·00	9·50	4·0	0·89
1 00	9·85	9·35	3·85	0·86
1 30	9·75	9·25	3·75	0·83
2 00	9·65	9·15	3·65	0·81
2 30	9·55	9·05	3·55	0·79
3 00	9·50	9·00	3·50	0·78
3 30	9·40	8·90	3·40	0·76
4 00	9·30	8·80	3·30	0·73
4 30	9·25	8·75	3·25	0·72
5 00	9·15	8·65	3·15	0·70
6 00	9·05	8·55	3·05	0·68
7 00	8·90	8·40	2·90	0·64
8 00	8·75	8·25	2·75	0·61
9 00	8·60	8·10	2·60	0·58
10 00	8·50	8·00	2·50	0·56
15 00	8·00	7·50	2·00	0·44
20 00	7·65	7·15	1·65	0·37
25 00	7·30	6·80	1·30	0·29
30 00	7·00	6·50	1·00	0·22
40 00	6·65	6·15	0·65	0·14
50 00	6·55	6·05	0·55	0·12
60 00	6·50	6·00	0·50	0·11

Notes: 1. Check whether a piezometer is required to establish true ground-water level.
2. During test measure the depth to water from top of casing.
3. Head (H) = $H_4 - H_3$ (falling-head test) and $H_3 - H_4$ (rising-head test).
4. At time $t = 0$, head $H = H_0$.

problems as the borehole should remain open during the test, after boring to the required level in advance of the casing. In softer clays it is possible that an unlined hole would collapse, but this can be prevented by installing a gravel filter over the test length. To do this the hole must be bored and cased to the required depth and the casing withdrawn after the gravel has been added. The hole must be carefully flushed out with clean water before the gravel is added, to ensure that the results of the test are not adversely affected by silting up.

In stiff fissured clay and some weak rocks the smearing action of the boring tool which tends to close fissures can be at least partly reversed by 'regenerating' the borehole by shelling out the water from the borehole prior to the test.

When tests are required in sands and gravels, they can normally be carried out with the casing flush with the bottom of the borehole. Careful 'shelling-out' will be necessary to avoid overboring the hole and the borehole must be kept topped up with water to prevent piping. Piping is likely to occur during a rising-head test in fine-grained materials and it is advisable to partially backfill the hole with about 1 m of gravel to counteract this. It will not be necessary to withdraw the casing unless the test is required over an unlined section of borehole. The hole should, if necessary, be flushed out so that the test is carried out in clean water conditions.

A careful check must be kept on the total length of casing when permeability tests are to be carried out in a borehole so that the position of the bottom of the casing can be accurately determined. Errors in this respect will lead to gross errors in the measured permeability.

The results can be seriously affected by bad joints in the casing allowing ingress, or escape, of water.

The most common tests carried out are rising- and falling-head tests, and whenever possible a rising-head and a falling-head test should be carried out at each level and the permeability quoted as the average result for both tests.

Normally a period of 1 h is allowed for each test, but in coarser granular soils the test will be complete in a much shorter time. In cases of very permeable soils it may not be possible to raise the level of water in the borehole.

When tests are being carried out in silts and clays of very low permeability, the standard test, restricted to a period of 1 h, is not likely to produce a reliable result. It may be necessary to carry out the test over a much longer period or to consider the use of a small-bore riser pipe within the borehole to achieve a more rapid change in water level, i.e. a shorter response time.

The average permeability of a rock stratum is determined by water injection (packer) tests in an exploratory drill hole. The test is carried out by sealing off a length of uncased hole with either hydraulic or mechanical packers, and injecting water under pressure into this isolated test section (Fig. 3.15). The rate of flow of water over the test length is measured (usually through a flow meter) under a range of constant pressures and recorded (Table 3.4). The permeability is calculated from the flow–pressure curve, a typical example of which is given in Fig. 3.16.

Although most commonly used for testing rock formations, packer testing can be used on soils, and in these cases it is usual for the upper packer to be placed just inside the lower end of the borehole casing.

The test can be carried out as drilling proceeds using a single packer to seal off the test length at the base of the hole. The sequence would be to drill the hole, remove the core barrel, insert and seal the packer assembly, carry out the test, remove the packer and drill the hole deeper. Because of the interruption in drilling time that this method of testing involves it is often preferable to carry out the tests as a sequential operation on a completed hole. Two packers are then employed to seal off selected test lengths, and the tests are performed from the base of the hole upwards. Usually successive test lengths are arranged to give a 50% overlap.

Before tests are carried out the hole must be flushed out with clean water to ensure that all the sediment has been removed. Flushing should be carried out from the bottom of the borehole and the final depth of hole checked by measurement. In weak fissured rocks the fissures may have been infilled by smearing and this could

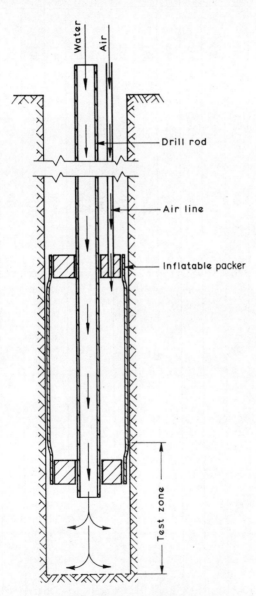

3.15. Drillhole 'packer' test arrangement

Table 3.4 WATER ABSORPTION TEST RESULTS

Water gauge pressure (kN/m²)	Pressure loss in drill rod and hose (kN/m²)	Static head gain (kN/m²)	Effective test pressure (kN/m²)	Duration of test (min)	Flow meter reading			Rate of flow (litre/min)
					start (litre)	finish (litre)	Total flow (litre)	
115	8	190	297	5	000	250	250	
				5	250	494	244	50·5
				5	494	758	264	
230	10	190	410	5	858	1160	302	
				5	1160	1465	305	60·0
				5	1465	1760	295	
345	14	190	521	5	1860	2215	355	
				5	2215	2548	333	69·5
				5	2548	2900	352	
230	12	190	408	5	3000	3329	329	
				5	3329	3650	321	67·0
				5	3650	4074	324	
115	10	190	295	5	4174	4478	304	
				5	4478	4775	297	60·5
				5	4775	5084	309	

Location No.: 8423
Carried out for: S & N Associates
Site: Deenham   Borehole No.: 32   Date: 12th April, 1974
Depth of hole at time of test: 20·0 m   Depth of casing: 10·0 m
Diameter in test area: 101 mm
Section tested by single/double packer from 18·5 m ... to 20·0 m below ground level

Borehole inclination and direction: —
Ground level: — (m, o.d.) Ground-water level: 5·0 m below ground level
Height of pressure gauge above ground level: at ground level
Length of hose between pressure gauge and drill rod: 3·0 m
I.D. of hose (nominal bore): 25 mm
Drill rod size: B.W. 'A' rod type

adversely affect the results of the tests. In such cases it is generally possible to clear the fissures with an inflow of water by initially reducing the piezometric head within the borehole. De-watering is carried out with a small compressor by lowering an open-ended airline to the base of the hole and allowing the air to force the water up out of the hole. De-watering should continue until the ejected water becomes clear. If a close examination of the recovered core shows that the fissures are normally clay filled, the hole should not be de-watered.

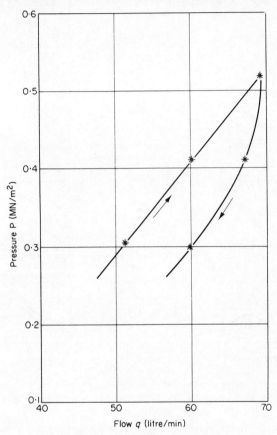

3.16. Flow-pressure curve

Specifications for the execution of packer tests generally require water to be pumped into the test section at a steady pressure, for periods of 15 min, with readings of the total water absorption being taken at 5 min intervals. The test normally includes five pumping-in cycles at varying applied water pressures. In metric units the successive pressures are normally specified as 6, 12, 18, then decreasing to 12 and 6 kN/m² per metre depth of packer below ground level. In Imperial units the test pressures are normally specified as 0·25, 0·5, 0·75, 0·5 and 0·25 lbf/in² per foot depth of packer below ground level. The required length of the test section will typically be 3·05 m (10 ft) or 6·1 m (20 ft).

The results should be recorded initially in tabular form (Table 3.4) and subsequently presented graphically to permit a qualitative assessment of the results to be made. For each test section P the applied pressure (ordinate), corrected for pipe friction losses, should be plotted against q the flow rate (abscissa). Some typical pressure P–q curves for tests in which the flow rate is recorded both for increasing and decreasing pressure are given in Fig. 3.17. The results may not plot as a

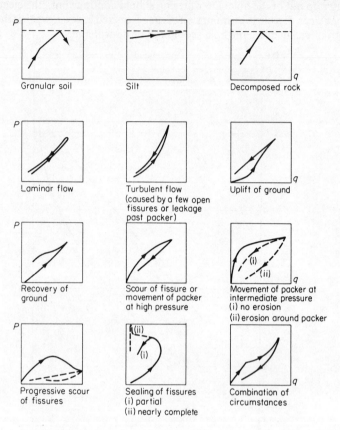

3.17. Typical pressure (P) against flow (q) curves for packer tests

straight line. As the pressure of water increases within the hole so the fissures may open causing a non-linear increase in flow rate. The final analysis of the results depends upon the position of the ground-water level. If this is not already known a piezometer must be installed so that the ground-water level can be accurately determined. The evaluation of the 'permeability' k from packer tests is normally based upon the methods proposed by the U.S. Bureau of Reclamation[10], using a relationship of the form

$$k = \frac{q}{C_\mathrm{S}RH}$$

where q is the steady flow rate under an effective applied head H (corrected for friction losses), R is the radius of the drill hole and C_S is a constant depending upon the length and diameter of the test section.

Frequently in cases where *in-situ* permeability tests are carried out to investigate the applicability of grouting methods the results of the tests are expressed in Lugeon units. The tests are conducted in a similar manner as before with measurements of water flow over a drillhole length of usually 5 m under an applied pressure difference of 1000 kN/m^2. The flow of water for these test standards is expressed in litres per minute per metre of hole, and these units are referred to as *Lugeons*. In grouting practice the suitability of different grout materials and injection techniques are often related to Lugeon units[11].

THE PRESSUREMETER

The Menard Pressuremeter is an *in-situ* testing device for testing both soils and rocks by expanding a probe in boreholes and drillholes of 50 or 75 mm diameter. The test is carried out where possible in an unlined borehole, but where necessary, a special slotted casing is employed to provide support, and the test carried out within this casing. The apparatus is shown diagrammatically in Fig. 3.18.

The probe is a cylindrical metal body over which are fitted three metal cylinders. A rubber membrane is stretched over these and clamped between them in such a manner as to form three independent cells. All three cells are inflated with water under a common gas pressure. Deformations are only measured by the central cell where conditions of plane strain are considered to exist due to the presence of the guard cells.

In order to minimise the possibility of damage to the cell membrane, the probe is enclosed in an outer rubber membrane which is generally protected by a shield of thin overlapping longitudinal metal strips.

The probes are available in various dimensions to suit standard borehole dimensions *AX, BX* and *NX*.

The volumeter consists of two identical metal water-filled reservoirs, equipped with pressure gauges and regulators. One reservoir is connected to the central measuring cell while the other is connected to the guard cells. They are linked to a common gas pressure supply and consequently water can be injected into the three cells at the same pressure.

The volume changes of the central cell are read directly from a graduated sight-tube indicating the water level in the measuring reservoir. For high modulus materials, the sight-tube may be isolated to give a reading sensitivity increased 50 times.

The volumeter is connected to the probe by a coaxial connecting line. The inner tube carries the water to the central cell while the annular tube feeds the guard cells. This prevents expansion of the measuring line under increasing pressure as a test proceeds.

Pressure is supplied from a commercial gas bottle. Various gases may be used, but nitrogen is to be preferred when higher pressures are required.

For each test the probe is inserted to the required depth at the end of a drill-rod assembly. A single pressuremeter test comprises at least 10 equal pressure increments with the related volume readings taken up to the ultimate failing pressure of the soil, in order that the pressure–volume curve can be properly delineated. In soft rocks the ultimate failing pressure may exceed the capacity of the apparatus ($9·8$ MN/m^2). Four volume readings are made at each pressure step at time intervals of 15, 30, 60 and 120 s after the pressure has become stabilised. It is customary to unload the soil at the end of the elastic phase of the expansion and to repeat the test before proceeding to the ultimate failing pressure.

After the appropriate corrections for the strength of the probe membrane, water-head imbalance and, where necessary, apparatus expansion under pressure, the results of each test are presented in the form of two graphs. One of these, termed the *PV* curve, relates pressure to volume change of the measuring cell. The other, termed the *creep* curve, relates the pressure to the increase in volume which occurred during the time interval 30–120 s at any particular pressure increment. The analysis of the test data will normally be carried out by the geotechnical specialist. The basic test results provide a value for the ultimate bearing capacity of the soil and also a value of the deformation modulus. The cohesion of clays can also be deduced from the test results.

While this test can be applied to any type of soil, it is particularly useful in the case of soils such as clays containing gravel. It is also very appropriate in the weathered zone of hard rocks and in weak rocks such as shales, marls and weak sandstones, which are unsuited to standard laboratory tests. As such they are an alternative to small plate bearing tests in boreholes. They are also useful as index tests where correlation with larger plate tests is possible, particularly in extending information to depth and below the water table. In granular materials, where the earlier part of the test is of less interest, the probe is sometimes driven either directly in loose deposits or within a slotted casing in denser materials.

Instrumentation

Geotechnical instrumentation of both soil and rock at ground surface and underground has now developed into a wide technology utilising a vast range of instruments and techniques. Most techniques are applicable in the measurement of the effects of underground excavation at the surface, although particular problems occur when working in areas of mining. Geotechnical instrumentation may be divided in general into the following categories:

1. Measurement of displacement or strain including tilt.
2. Measurement of stress or stress change.
3. Measurement of water flow and pressure.

The first of these is the most important in the measurement of the effects of mining subsidence. The techniques available in all categories are wide and in general the approach to instrumentation depends on the expected magnitudes of the effects to be measured. Surface effects of underground mining have been shown by Orchard[12] and others to include vertical settlements of the order of 0·8–0·9 times the subsurface closure, ground tilt of up to about 4° and surface horizontal strain of up to 6% (both tensile and compressive) although figures of 1% are more common. The actual movements likely in any given situation depend on the geometry of the mining and the geology and it should be remembered that surface evidence of subsidence, i.e. building cracking, etc. is often manifest only at a late stage of movement and therefore early measurements should be instituted, preferably prior to the commencement of mining operations. Surface effects will be measurable immediately following mining; tilt, displacement and in particular horizontal strain, will change as mining progresses and in the period following the completion of mining.

Typical examples only are described in all types of measurement as the range

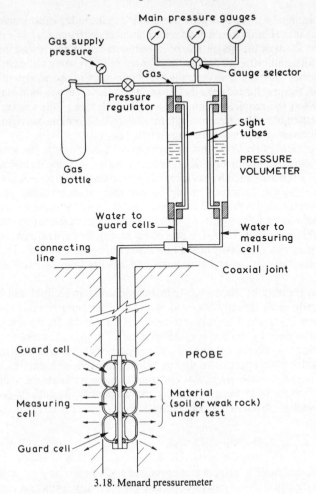

3.18. Menard pressuremeter

available is very wide, although the most suitable instruments for a given situation often provide only a narrow choice. Any decision must be made with consideration of the performance characteristics of the instrument and the parameter to be measured. As already stated, the most suitable time to install instruments is before any change in ground conditions is made, as this provides a reliable base reading. However, it is possible under certain circumstances to make useful measurements after an apparent problem has occurred during construction or excavation.

DISPLACEMENT MEASUREMENT

Methods of measuring surface or subsurface movement or strain vary from conventional surveying techniques to sophisticated electronic equipment based on electrical spirit levels, accelerometers, etc. In order to measure the total magnitude of movement the measurements must be related to a remote datum which is known to be stable and it is this which provides a difficulty in mining subsidence as the surface effects often obtain over a wide area. The effect on structures however is not so much a problem of the magnitude of movements but of differential movements, tilts and horizontal strain. There are many methods of relating settlement to a central point, but the most convenient is the pneumatically operated settlement pot which uses the pressure change due to a column of mercury to indicate change in level between the required point and a central 'datum'. This system indicates relative settlement only unless the datum is related to a remote bench mark which may be done using precise surveying techniques.

Displacement of the ground with depth may be measured in the vertical sense using borehole extensometers and in the horizontal sense using inclinometers. For the former a series of anchors at various depths in a borehole may be connected to the surface using wires or rods and the displacement of each anchor relative to the surface measured. These in fact may also be used underground to measure displacement at any angle towards an opening or excavation. The inclinometer relies on the measurement of the angle a probe makes with the vertical as it is lowered or raised in stages in a specially cased borehole and, because of basic inaccuracies, should only be used where horizontal movements greater than 50 mm are expected.

Ground tilt may be measured by installation of a spirit level which is either mechanically or electrically operated, with the latter being the most convenient and accurate. A ground-tilt measurement system suitable for mining settlement has been described by Whittaker and Forrester[13]. They also describe a mechanical strain-measuring device suitable for estimating horizontal ground strain, which uses a base length of approximately 3 m. Where more precise estimates of horizontal strain are required, in particular on structures and foundations, shorter gauge lengths of about 0·25 m are probably more suitable. Results of this type of measurement are discussed by Littlejohn[14].

STRESS AND STRESS CHANGE

The measurement of stress, contact pressures and stress change may be made in two ways. Strain may be measured in any one of many ways and converted to stress by the assumption of soil or rock parameters. However, by the inclusion of a stiff measuring device, stress changes may be measured directly. An example of this type of instrument is the Glotzl cell, described by Haws *et al.*[15], which is a hydraulic cell having a high stiffness at constant temperature and is used for measuring contact pressures. Borehole rigid inclusion cells are used for installation underground to measure stress changes as a result of mining or tunnelling.

WATER FLOW AND PRESSURE

Many techniques are available for estimating velocities and directions of underground water flow, although each has particular limitations. The flow direction of water can be determined in a number of ways, but two techniques which are commonly used involve the monitoring of tracers, either dyes or radioactive sources, which are introduced into the ground-water generally through boreholes. The piezometer, a simple form of which has been described earlier in the chapter, is used to measure water pressure and is available in several forms with the alternative of hydraulic, pneumatic or electrical read-out systems. The hydraulic system is in most cases the best system as de-airing of the tip at any stage is possible although the read-out system is probably least convenient to handle. Development, for providing a de-airing facility for the pneumatic and electrical systems is currently taking place. As emphasised in the section dealing with ground-water, installation of all piezometer types must be carefully carried out as their performance and measurements will be seriously affected by faults during installation.

Conclusions

The techniques of field investigation have been set out in this chapter with an indication of how the quantitative data can be applied to certain problems. It is obvious that in some subsidence problems extensive and sometimes special laboratory testing of both soils and rocks will be necessary. However, laboratory testing is outside the scope of this chapter. The points to be emphasised regarding field investigation may be summarised as follows:

1. Before embarking on any field work seek out existing geological data.
2. Extend the existing data by a reconnaissance survey.
3. Plan and carry out the field work programme modifying it in the light of the data which become available as the work proceeds.
4. Pay particular attention to the obtaining of correct piezometric data.

It is important that any exploration programme should be supervised by an experienced specialist engineer or engineering geologist with several years experience of geotechnical problems. The description of samples and cores requires a certain degree of interpretation, and although this can be competently carried out by a supervised graduate engineer or geologist of modest experience, occasional logs should be checked by re-examination and description carried out by the senior man. Constant supervision is also required of the boring/drilling operators to ensure that the wrong procedures such as overshelling, allowing piping to occur, poor or inaccurate recording of water levels, etc. are not allowed to creep in. However carefully and accurately the data obtained are applied to the problem it is to no avail if the basic data themselves are inaccurate.

REFERENCES

1. Dumbleton, M. J. and West, G., *Preliminary Sources of Information for Site Investigations in Britain*, RRL Report LR403, Dept. of Environment, Crowthorne (1970)
2. Working Party Report on the Logging of Rock Cores, Engineering Group of the Geological Society of London (1970)
3. *Site Investigations:* British Standard Code of Practice CP 2001: British Standards Institution, London (1957)
4. Tomlinson, M. J., *Foundation Design and Construction*, Pitman, London (1969)
5. Sanglerat, G., *The Penetrometer and Soil Exploration*, Elsevier Book Division, Paris (1972)
6. *Symposium on Vane Shear Testing*, ASTM Special Tech. Publication No. 193, 5 papers and discussion (1956)
7. Skempton, A. W., 'Vane Tests in the Alluvial Plain of the River Forth, near Grangemouth', *Geotechnique*, **1**, 111 (1948)
8. Plaate, K., 'Factors Influencing the Results of Vane Tests', *Canadian Geotechnical Journal*, **13**, 18 (1966)

9. Hvorslev, M. J., *Time Lag and Soil Permeability in Groundwater Observations*, Bulletin No. 36, Waterways Experimental Stations, Vicksburg (1951)
10. *Earth Manual*, US Bureau of Reclamation, 544–546, 1st edn (1968)
11. Cambefort, H., *Forages et Sondages*, Eyrolles, Paris (1959)
12. Orchard, R. J., 'Surface Effects of Mining—the Main Factors,' *Colliery Guardian*, 193 (1956)
13. Whittaker, B. N. and Forrester, D. J., 'Measurement of Ground Strain and Tilt Arising from Mining Subsidence', in *Field Instrumentation in Geotechnical Engineering*, Butterworths (1974)
14. Littlejohn, G. S., 'Observation of Brick Walls Subjected to Mining Subsidence.' *Symp. on Settlement of Structures*, Pentech Press (1974)
15. Haws, E. T., Lippard, D. C., Tabb, P. and Burland, J. B., 'Foundation Instrumentation for the National Westminster Bank Tower', in *Field Instrumentation in Geotechnical Engineering*, Butterworths (1974)

BIBLIOGRAPHY

Golder, H. A. and Gass, A. A., *Field Tests for Determining Permeability of Soil Strata*, ASTM Spec. Tech. Pub. No. 322, 29 (1963)
Muir Wood, A. M. and Caste, G., 'In-situ Testing for the Channel Tunnel', *Proc. In-situ Investigations in Soils and Rocks*, 1969, British Geotech. Soc., 109 (1970)
Terzaghi, K. and Peck, R. B., *Soil Mechanics in Engineering Practice*, John Wiley and Sons, New York, 326 et seq. (1967)
A New Approach for Taking a Continuous Soil Sample, Laboratorium voor Grondmechanica, Paper No. 4, Delft, Holland (1966)
Permeability and Capillarity of Soils, ASTM (1967). This collection of papers includes an extensive bibliography

Chapter 4

Geophysical Methods in Site Investigations in Areas of Mining Subsidence

Many engineers are doubtful about the value of indirect or geophysical methods in site investigation. The present authors' view is that these methods are a useful site investigation tool provided they are used in the right context. To some extent they complement the direct or physical methods—drilling and test pitting—in that the latter give definite information at the points at which they are excavated, but tell nothing about the ground between, whereas the geophysical methods are well adapted to detecting variations in subsurface conditions between boreholes. In this respect, therefore, an appropriate combination of direct and indirect methods would result in a better standard of site investigation than is often the case at the present time.

In deciding whether to employ geophysical methods on a particular site investigation, the geological conditions on the site need to be sufficiently understood to answer the following questions:

1. Do the principal geological formations differ in one or other physical property sufficiently to permit them to be distinguished by physical measurements made at the ground surface?
2. Are the geological conditions sufficiently close to the relatively simple models used in developing the theoretical bases of the geophysical methods to enable the theory to be applied?

The character and situation of the site also have to be taken into account. Sites in built-up areas may be unsuitable for one or other of the geophysical methods, either because of old construction or services on the site, interference from some source, or lack of space for carrying out the survey. Sites in open country are generally suitable although interference may occur in the vicinity of power lines, heavily trafficked roads, etc.

These two questions can be answered satisfactorily only by those with experience in this type of work. Too often in the past geophysical methods have been used without due reference to the geological situation, and consequently the results have been disappointing. These failures always appear to be blamed on the method rather than on the mis-application of the method, and this is why many engineers mistrust geophysical methods. The solution to this difficulty is to take better geological advice during the planning of the investigation, and to maintain close geological supervision during its execution.

If conditions are favourable for a geophysical survey, and if the information such a survey would provide is required for the design of the project, then a geophysicist should be consulted to plan the survey. The plan should be based on the full knowledge of the geology, and on a clear definition from either the geologist or engineer of what the survey is expected to achieve.

Definition of Objectives

Introduction   The objectives of a geophysical survey forming part of a site investigation in a mining area are likely to be one or more of the following:

1. Detection of old mine shafts including bell-pits.
2. Detection of old underground workings.
3. Location of zones of faulting in the Coal Measures strata along which subsidence due to collapse of old workings could be expected to be concentrated.

Old Mine Shafts   In considering old mine shafts as a target for detection by geophysical methods, account should be taken of the possible variations, both in original construction and later history, that may be present. The principal variable factors are as follows:

Original construction
Unlined   Brick lined

Later history
Infilled: Open
    complete; partial (rafted over   (rafted over at ground surface)
    at depth)

    Nature of infilling:
        material from original excavation;   air above water table,
        mine waste mixed with debris from   water below water table
        mine buildings, headworks, etc.

Thus the target for the geophysical survey may vary from an unlined bell-pit of 1–2 m in diameter and up to about 13 m deep, completely backfilled with the material originally excavated from it, to an open brick-lined shaft of 2 m diameter, partly air filled and partly water filled.

Old Workings   Old workings are also quite variable, depending on:

(*a*) Thickness of the seam.
(*b*) Percentage of coal extracted in room and pillar, or pillar and stall workings.
(*c*) State of preservation or collapse of old workings.
(*d*) Level of the water table.

Thus the target for the geophysical survey may vary from a water or air-filled cavity of 1–2 m in height, to a zone of collapsed strata several times that height.

Fault Zones   The importance of faults 'as safety valves . . . absorbing strains in an area' has been pointed out by Taylor[1]. Briggs[2] and a publication[3] on mining subsidence recommend a stand-off of 15 m either side of known surface positions of faults. In cases where the precise position of a fault is unknown due to poor exposure resulting from a cover of glacial drift or other overburden, geophysical methods can be used to locate the fault, provided one of the following conditions is satisfied:

1. There is a measurable contrast in one or other physical property between the two formations on opposite sides of the fault.
2. There is a zone of fractured, broken rock associated with the fault, wide enough to permit detection.

Correlation Boreholes   Boreholes to provide correlation for the geophysical measurements are an essential part of any geophysical survey. This is especially true of surveys for old mine workings in view of the possible variations in the target already outlined. In the theoretical case mentioned by Taylor[1] of boreholes put down on a grid coinciding with pillars of coal if the old workings form a regular pattern, the objective of a geophysical survey would be to detect the change in ground conditions between the boreholes caused by the old workings, and thereby point to the need for additional borings. This procedure of sinking a number of boreholes prior to the start of the geophysical survey, followed by additional borings sited on the basis of the survey, is considered to have general application in all site investigations.

Geophysical Methods and Theory

INTRODUCTION

The four main geophysical properties that have been used (either separately or in some combination) in searches for the after effects of mineral extraction are listed in Table 4.1. The methods are listed in approximately decreasing order of importance in so far as the present objective is concerned, and each is discussed in some detail subsequently.

Table **4.1** PHYSICAL PROPERTIES AND GEOPHYSICAL METHODS

Physical property	Geophysical method
Electrical resistivity	Resistivity
Density	Gravity
Magnetisation	Magnetics
Seismic-wave velocity	Seismic refraction and/or reflection

In these discussions the instrumentation may appear to have been neglected. This is because the authors consider that it is more important for the engineer to understand the theoretical bases of the methods, the relatively simple earth models that can be deduced from the theories and thus the inherent limitations of the methods, than it is for him to know how the measurements are made.

THE ELECTRICAL RESISTIVITY METHOD

The electrical resistivity of a uniform cylindrical piece of material of length l and cross-sectional area a is defined by:

$$\rho = \frac{Ra}{l} \qquad (4.1)$$

where R is the ohmic resistance of 1 cm^3 of the material. The S.I. unit is the ohm/metre (Ωm).

Rocks and minerals exhibit resistivities ranging from 10^{-3} to $10\ \Omega$m, and Table 4.2 indicates the resistivity values of several important rock types. Generally, the rock-forming minerals are poor conductors (i.e. have high resistivities) and the resistivity of rock masses depends largely on the pore space, the water filling the pores and the presence of dissolved salts in these ground-waters. Conversely, if the pores or other voids within a rock mass are air filled, the resistivity will be increased still further. In the native metal ores metallic conduction occurs; sulphide ores and carbonaceous deposits are semiconductors; and the rock-forming silicates are solid electrolytes. Particularly important in the last group are the common clay minerals whose crystal lattices have many ions readily available for electrolytic conduction.

Table 4.2 RESISTIVITIES OF ROCKS AND MINERALS

Rock or mineral	Resistivity (Ωm)
Igneous rocks	
granite	300–10 000
diorite	50 000
gabbro	100–15 000
dolerite	20–20 000
basalt	100–20 000
Metamorphic rocks	
gneiss	200–20 000
schists	5–10 000
slates	1–500
Sedimentary rocks	
coal	1–200 000
sandstone	30–100 000
limestone	60–500 000
shale	8–100 000
Minerals	
quartz	$> 100\ 000$
calcite	$> 100\ 000$
graphite	8×10^{-6} to 6×10^{-2}
diamond	1×10^{12}
clay minerals	10–30

Considering the potential distribution within a homogeneous earth of resistivity ρ arising from a current source at the surface, if the current flow into the earth at the source is I, then the potential difference dV across a hemispherical shell of radius r

and thickness dr will be:

$$\mathrm{d}V = \frac{-I\rho\,\mathrm{d}r}{2\pi r^2} \tag{4.2}$$

By integration the potential V at a distance r from the source can be obtained:

$$V = \frac{I\rho}{2\pi} \times \frac{1}{r} \tag{4.3}$$

In resistivity surveys four electrodes are commonly used; two are the current source and sink (often denoted by A and B) and two potential electrodes (denoted by M and N) are inserted in various collinear configurations. Some examples of these configurations are shown in Fig. 4.1. The expression for the potential difference between the potential electrodes is obtained by algebraic summation of the potentials due to the two current electrodes at the two potential electrodes.

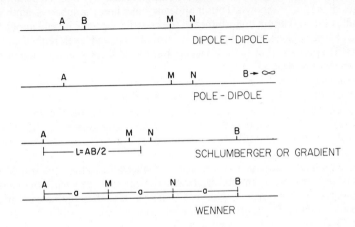

4.1. Resistivity electrode array configurations

Hence equation 4.3 becomes:

$$\rho = \frac{2\pi\Delta V}{IG} \tag{4.4}$$

where G is a geometrical function relating to the array configuration, i.e.

$$G = \frac{1}{AM} - \frac{1}{BM} - \frac{1}{AN} + \frac{1}{BN} \tag{4.5}$$

Because the ground is inhomogeneous, ρ is not constant everywhere and in practice an apparent resistivity ρ_a is measured which is a 'bulk' value for the sub-surface materials near the electrode array.

Evidently, as the current electrode separation AB increases (Fig. 4.1), so the depth penetrated by the current increases. Consequently, for small values of AB, ρ_a closely approximates to the resistivity of the uppermost layer, while for larger AB values ρ_a is more dependent upon the resistivity of deeper layers.

Two types of measurement are commonly employed in resistivity surveys. In one type, known variously as *depth probes, electrical sounding,* or *electrical drilling,* an expanding electrode array centred on one point effectively measures the sequence of horizontal layers of contrasting resistivities below the survey point. In the second method an electrode array of fixed dimensions is moved across the survey area and detects variations in resistivity within a maximum depth dependent upon the array dimensions. This technique is known variously as *resistivity* (or *electrical*) *traversing,* or as *electrical trenching.* Examples of the results of both types of measurement are shown in Fig. 4.2.

Horizontal resistivity profiles (using the Wenner array) across a buried perfectly conducting sphere, taken from Van Nostrand and Cook[4] are shown in Fig. 4.3. These curves were obtained from the solution of Laplace's equation in bipolar co-ordinates, and show the variation of the ratio of apparent resistivity to true resistivity of the country rock against the ratio of electrode separation to sphere depth. A flooded spherical cavity could be expected to give similar curves, whereas the curves due to an air-filled cavity, with a high resistivity, would resemble mirror images of the curves shown. Thus it appears that where the sphere's depth is twice its radius the maximum disturbance in the resistivity profile is only about 10% of the background level. Since noise about the background level frequently exceeds 10% it is apparent that bodies deeper than twice their average dimension will probably not be detectable.

In the case of the horizontal buried cylinder (Fig. 4.4) it can be seen that the cylinder is detectable at slightly greater depths than the sphere.

The decrease of the anomaly due to a buried sphere as the distance of the traverse from the epicentre of the sphere increases, is shown in Fig. 4.5. This effect is not relevant in the case of the cylinder which is assumed to be infinite in the direction normal to the plane of the diagram.

The practical consequences of these considerations affect the direction and separation of traverses, and the electrode separations to be used. For example, in searching for approximately spherical bodies of, say, 10 m diameter centred at a depth of 6 m with a resistivity about one fifth that of the surrounding country rock, Fig. 4.2 indicates that the maximum anomaly occurs with an electrode separation of about 0·9 times the body radius (i.e. 4·5 m). Furthermore, the curves of Fig. 4.4 show that the separation of the traverses should not exceed the approximate average dimension of the sought-after body.

Together, these two parameters, i.e. depth and size, impose quite severe restrictions on the resolution of the resistivity method in any search for small bodies such as old mine workings, shafts and bell-pits. Several case histories of resistivity surveys are referred to later.

In conclusion, it is worth emphasising that while it is theoretically feasible to detect cavities (including old mine workings) by resistivity measurements, in practice the depths at which these commonly occur are too great in relation to the size of the working to produce anomalies distinguishable from 'noise' generated by other near-surface electrical inhomogeneities.

THE GRAVITY METHOD

The gravitational acceleration caused by the Earth's mass is approximately 9780·49 mm/s^2 at the equator and increases gradually by about 50 mm/s^2 towards

4.2. Examples of data yielded by vertical electrical
sounding and by resistivity traversing

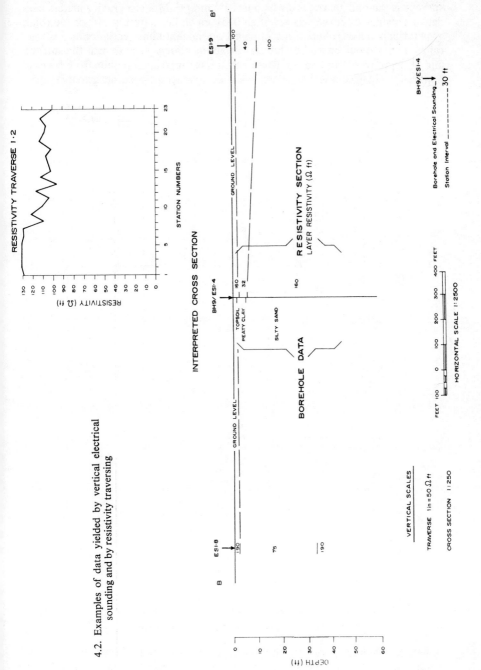

the north and south poles. The unit of acceleration most commonly used in gravity surveys is the milligal (mgal), which is 10^{-2} mm/s². Modern gravity meters can detect changes in gravity as small as 0·01 mgal, i.e. 1 part in 10^8 of the total gravitational acceleration. Superimposed on the latitudinal variation is a whole range of variations caused by laterla variations of density both near the surface and at depth within the Earth. These variations range in amplitude from several hundred milligals to less than can be resolved by the most accurate gravity meter.

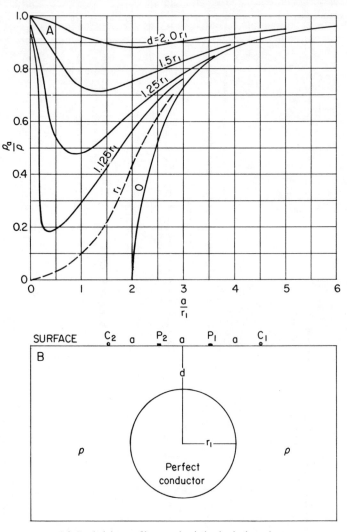

4.3. Resistivity profiles over buried spherical conductor

Before considering further the nature of the variations in gravity, usually referred to as 'gravity anomalies', it is necessary to discuss the densities commonly found in rocks. Table 4.3 contains a list of rock densities; evidently for rocks generally encountered at mineable depths they range from about 1·5 g/cm³ to

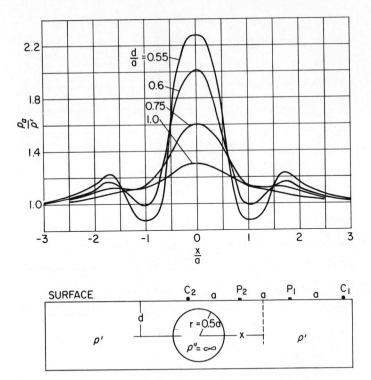

4.4. Resistivity profiles over buried horizontal cylindrical insulator

Table 4.3 SOME TYPICAL ROCK AND MINERAL DENSITIES

Rocks	Source reference	No. of specimens	Mean density (g/cm³)	Range of density (g/cm³)
Igneous				
granite	1	155	2·667	2·516–2·809
diorite	1	13	2·839	2·721–2·906
gabbro	1	27	2·976	2·850–3·120
rhyolite	2	—	2·55	2·4–2·7
basalt	2	—	3·0	2·7–3·3
Metamorphic				
acid gneiss	3	59	2·72	2·64–2·78
basic gneiss	3	20	2·97	2·8–3·1
schists	3	14	2·83	2·76–2·90
slate	2	—	2·8	2·7–2·9
Sedimentary				
coal	4	—	1·5	1·2–1·8
sandstone	4	—	2·3	1·8–2·7
limestone	4	—	2·65	2·6–2·7
shale	4	—	2·2	1·5–2·8
chalk	4	—	2·0	1·5–2·4

Source references: 1 Clarke (1966), 2 Jakosky (1950), 3 Maton (1971) and 4 Parasnis (1972).

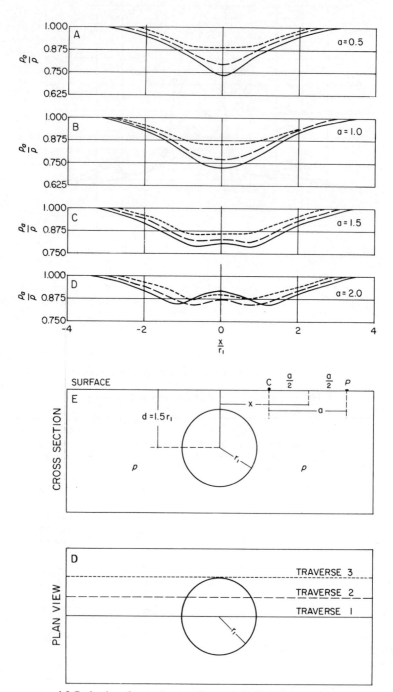

4.5. Reduction of anomaly away from epicentre of a buried sphere

about $3 \cdot 2$ g/cm³. From the point of view of gravimetry, the important parameter is the density contrast between the anomalous mass sought by the gravity survey and its enclosing country rock. In the case of underground workings, the contrast is invariably negative, but its magnitude can vary according to whether the workings have been backfilled (and, if so, how completely), or whether they are flooded or air filled. Thus, in average Coal Measures where the formation density would be about $2 \cdot 7$ g/cm³, if old workings were 70% effectively backfilled the fill would have a density of approximately $1 \cdot 9$ g/cm³, and a contrast of $-0 \cdot 8$ g/cm³ would result. For flooded and open workings the contrasts would be $-1 \cdot 7$ g/cm³ and $-2 \cdot 7$ g/cm³, respectively.

The fundamental law of gravitation is, of course, Newton's law which in effect says that the acceleration a caused by a point mass m at a distance r from m is:

$$a = \frac{Gm}{r^2} \tag{4.6}$$

where G is the Newtonian constant and has the value of $6 \cdot 67 \times 10^{-11}$ SI units. In practice geological bodies have finite volumes and more complicated expressions apply. Some simple bodies, to which old excavations may be assumed to approximate, are depicted in Fig. 4.6 together with the associated two-dimensional anomaly along profiles perpendicular to the axes of the bodies. The corresponding expressions are written out below, using the notation of Fig. 4.6:

$$\textit{Sphere} \quad \Delta g = \tfrac{4}{3} \pi R^3 G \Delta\rho \, \frac{z}{(x^2 + z^2)^{3/2}} \tag{4.7}$$

$$\textit{Horizontal cylinder} \quad \Delta g = 2\pi R^2 G \Delta\rho \, \frac{z}{(x^2 + z^2)} \tag{4.8}$$

$$\textit{Vertical cylinder} \quad \Delta g = 2\pi G \Delta\rho [z_2 - z_1 + (z_1{}^2 + R^2)^{\frac{1}{2}}$$
$$- (z_2{}^2 + R^2)^{\frac{1}{2}}] \tag{4.9}$$

(This applies only on the axis of the cylinder; off the axis, elliptic integrals are required to evaluate Δg.)

$$\textit{Vertical prism} \quad \Delta g = 2G \Delta\rho \left[x \ln \frac{r_1 r_4}{r_2 r_3} + l \ln \frac{r_1}{r_2} + z_2(\phi_2 - \phi_1) \right.$$
$$\left. - z_1(\phi_4 - \phi_3) \right] \tag{4.10}$$

If the vertical faces are moved laterally to infinity, $r_1 \to r_2, r_3 \to r_4, \phi_1$ and $\phi_2 \to \pi$, and ϕ_3 and $\phi_4 \to 0$; then

$$\Delta g = 2\pi G \Delta\rho(z_2 - z_1) \tag{4.11}$$

This is the expression for the acceleration caused by an infinite horizontal slab.

It is instructive to insert values for the variables in some of these expressions to indicate the likely gravity anomalies caused by old mine workings. An air-filled horizontal tunnel of cylindrical cross-section, e.g. 5 m in radius and 20 m below ground level, would produce an anomaly of $0 \cdot 024$ mgal. A gallery 2 m high, air filled, and approximated by a semi-infinite horizontal slab would produce a maximum anomaly of $0 \cdot 2$ mgal.

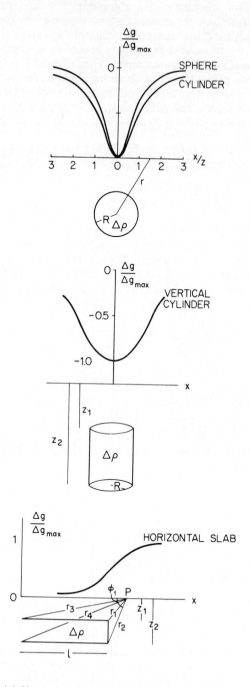

4.6. Simple shaped bodies and their gravity anomalies

While gravity meters are in theory capable of resolving such small anomalies, in practice it is often difficult to eliminate other anomalies of similar magnitude caused by near-surface density variations, when applying the necessary corrections for topography and other effects to the raw gravity field data. Colley[5] discussed the use of gravimetry in the detection of subterranean caves in some detail. The model calculations and data presented by Colley can be summarised as follows: for a limit of resolution of 0·2 mgal, the ratio of maximum cavity dimension to the depth to the top of the cavity was never less than 1 : 4. In mining areas, the cavities are both generally smaller than natural caves, and occur at greater depths. It follows therefore that gravity survey is not a powerful tool in the search for old mine workings. An exception to this would be in determining the boundaries of backfilled open-cast workings which have been obscured subsequently. In such cases the density contrasts, though smaller than for empty or flooded underground cavities, are near or at the ground surface and often have greater vertical extent than the underground excavations.

THE MAGNETIC METHOD

The magnetic method detects variations in the geomagnetic field strength (or one of its vector components). These variations are caused by contrasts in the magnetisation of different rocks, and depend largely on variations in their content of ferromagnetic magnetite-type minerals. Magnetisation is a vector quantity, unlike density, and therefore can contain several components. The two main components of magnetisation are, firstly, the induced magnetisation J caused by the present Earth's field H ($J = KH$ where K is the susceptibility) and secondly, the remanent magnetisation usually acquired at the time of formation of the rock.

The Earth's magnetic field varies in strength from about $0·38 \times 10^{-11}$ Wb/m^2 near the tropics to about $0·65 \times 10^{-4}$ Wb/m^2 near the magnetic poles. The unit of field strength most used in geophysical surveys is the gamma (γ), where $1\gamma = 10^{-9}$ Wb/m^2. Nuclear resonance magnetometers have sensitivities of $0·1–0·05\gamma$, while fluxgate and proton precession magnetometers have an accuracy of 1 or 2γ.

Some typical rock susceptibilities are listed in Table 4.4. Evidently, basic igneous and other crystalline rocks have the greatest susceptibilities. However, mining excavations in these rocks for metallic ores generally will not produce bodies (voids) of sufficient magnitude to generate a resolvable magnetic anomaly.

The magnetic method has proved useful, however, in detecting old mine workings in which either brick linings or metallic objects such as rails or casing have been abandoned. Bricks generally contain a significant quantity of magnetite or haematite and have both a quite strong thermal remanent magnetisation[6] and an appreciable induced magnetisation. Since the bricks in any structure are quite disorientated from the position in which they were baked and hence acquired their remanent magnetisation, it follows that any anomaly generated by them in the mineshaft structure will be due to their induced magnetisation.

The form of anomaly over a vertical cylinder in which the magnetisation of the cylinder is directed vertically downwards and is uniform, is shown in Fig. 4.7. This latter condition is somewhat idealised and is unlikely to occur in real situations. However, anomalies ranging in amplitude from a few gammas to several hundreds of gammas or more have been found associated with mineshafts.

Table 4.4 SOME ROCK SUSCEPTIBILITIES

Rock types	Range of susceptibilities (rationalised SI units × 10⁶)
Igneous	
granites	380–34 000
gabbro	5 520–51 500
dolerite	620–52 700
basalt	8 550–79 100
Metamorphic	
gneiss	120–25 000
schist	330–3 000
slate	490–37 700
serpentine	3 140–176 000
Sedimentary	
sandstones	63–1 260
shales	approx. 500
clay	approx. 250
carbonates	11–50

Note: to correct the above values to unrationalised e.m./c.g.s. units, divide by 4π.

MAGNETIC INCLINATION = 60°

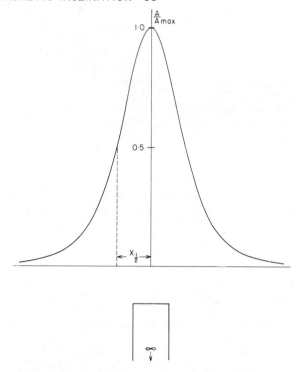

4.7. East to west magnetic anomalies over a vertical magnetised cylinder

The interpretation of anomalies over mineshafts is qualitative. When an anomaly of amplitude significantly greater than the local background noise level is discovered, the position is marked and subsequently either drilled or trenched. The latter method has the advantage that fill material is more easily recognised. Instances occur of old shafts being rafted at depth and backfilled to the surface. In such cases the fill may modify the anomaly considerably. The local background level is another critical factor.

In the Fife coalfield a magnetic survey was undertaken for old mineshafts and several anomalies of up to a thousand gammas were discovered. In fact so many anomalies were recorded that old mining records were researched. These indicated that the survey area had been the spoil tip for ironstones extracted simultaneously with the coals, and the ironstones were causing the anomalies. Fences, drains and power lines are sources of interference more usually encountered in developed areas.

There are several 'rule of thumb methods' applicable to anomalies in potential fields which enable estimates to be made of the depth to the top of the anomalous mass or magnetisation. One of these, indicated in Fig. 4.7 is that the depth d is approximately twice the half width of the anomaly at the half maximum amplitude of the anomaly; in the notation of Fig. 4.7:

$$d \approx 2x_{\frac{1}{2}} \qquad (4.12)$$

Such estimates may be of use in deciding the subsequent method of physical examination. They should be regarded, however, as gross approximations.

THE SEISMIC METHODS

Rocks normally propagate two types of shock wave, i.e. shear S waves and longitudinal P waves. The relationships between the velocities of shear and longitudinal waves (denoted V_s and V_p respectively) and the elastic parameters of isotropic media are:

$$V_s = \left(\frac{\eta}{\rho}\right)^{\frac{1}{2}} \text{ and } V_p = \left(\frac{K + \frac{4}{3}\eta}{\rho}\right)^{\frac{1}{2}} \qquad (4.13)$$

Where η = Shear modulus,

  K = Bulk modulus and

  ρ = Density.

Clearly, V_p is greater than V_s and thus P-waves are usually the first to arrive at a seismic detector, and with most recording equipments currently used in site investigations, they mask subsequent arrivals.

Since shear waves will not be transmitted by fluids ($\eta = 0$), the detection (or non-detection) of shear waves could be of great value in studies of air- and water-filled underground cavities.

Both the refraction and reflection techniques measure the thicknesses and seismic wave velocities of fairly uniform near-horizontal layers of rock. Clearly, underground excavations do not often conform with the above simple model, and thus seismic methods have not been extensively used in searches for underground cavities of whatever origin.

Interpretation of seismic data rests fundamentally on the principles of geometrical optics. Huyghen's principle of the propagation of wavefronts is assumed, as is Fermat's principle of minimum time paths. Snell's laws of refraction and reflection are used to evaluate time–depth relationships and the derivations of these can be summarised as follows:

Assuming two media of seismic velocities V_1 and V_2, in which V_1 overlies V_2, Snell's law of reflection states (Fig. 4.8a) that the angle of incidence i equals the angle of reflection r.

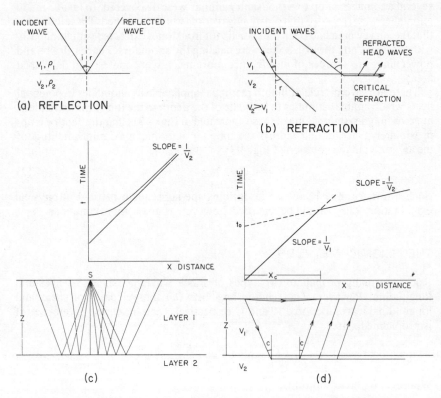

4.8. Seismic reflections (*a*), refractions (*b*), and time–distance graphs (*c* and *d*)

In the simple situation of Fig. 4.8c, the total time for a sound wave to travel from the source S to the receiver R is:

$$t = \frac{2}{V_1}\left(\frac{x^2}{4} + z^2\right) \qquad (4.14)$$

$$\therefore \qquad z = \tfrac{1}{2}(V_1^2 t^2 - x^2)^{\frac{1}{2}} \qquad (4.15)$$

V is obtained either from the slope of a plot of t against x for large values of x, or by the slope V^{-2} of a plot of t against x^2, since the times t increase or 'step out' with the angle of incidence.

In the case of seismic refraction (Fig. 4.8b) where $V_2 > V_1$ refraction in Layer 2 is towards the horizontal and Snell's law states that:

$$\frac{\sin i}{\sin r} = \frac{V_1}{V_2} \tag{4.16}$$

For the case of critical refraction, $\sin r = 1$ and $i = c$, the critical angle. The critically refracted wave in Layer 2 itself propagates head waves upwards from the boundary at the critical angle through Layer 1. For such waves the travel time t is derived as follows (Fig. 4.8d):

$$t = \frac{2z}{V_1 \cos c} + \frac{x - 2z\tan c}{V_2} \tag{4.17}$$

and

$$t = \frac{2z}{\cos c}\left(\frac{1}{V_1} - \frac{\sin c}{V_2}\right) + \frac{x}{V_2} \tag{4.18}$$

From Snell's law

$$\sin c = \frac{V_1}{V_2}$$

and

$$\cos c = \left(1 - \frac{V_1^2}{V_2^2}\right)^{\frac{1}{2}}$$

Substituting into equation (4.18) gives:

$$t = 2z\frac{(V_2^2 - V_1^2)^{\frac{1}{2}}}{V_1 V_2} + \frac{x}{V_2} \tag{4.19}$$

A graph of t against x thus has slope $1/V_1$ to the critical distance x_c given by:

$$x_c = 2z\left(\frac{V_2 + V_1}{V_2 - V_1}\right)^{\frac{1}{2}} \tag{4.20}$$

and slope $1/V_2$ beyond x_c. Also, t_o for $x = 0$ is given by:

$$t_o = 2z\left(\frac{V_2^2 - V_1^2}{V_1 V_2}\right)^{\frac{1}{2}} \tag{4.21}$$

Similar expressions can be obtained for cases of more than one refracting boundary.

A situation in which reflection and refraction of seismic waves are affected by cavities, i.e. in which V_p is reduced and V_s is zero, is shown in Fig. 4.9. Refracted waves are incident on the cavity in Layer 2 from the left-hand side of the diagram and reflected waves are incident from the right-hand side. Also shown are the time–distance graph of the refracted wave arrivals and wiggly line representations of the reflected arrivals. The following discussion of the reflections is based on the paper by Cook[7].

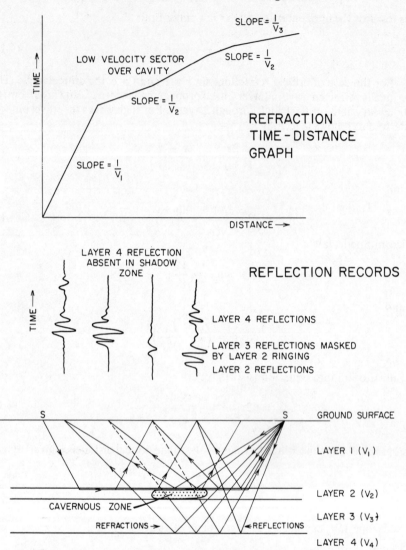

4.9. Idealised seismic effects around a cavity

Seismic energy incident on a boundary between two media is split into several parts; some is reflected, some refracted, and some converted into other wave types. For near-vertical incidence of approximately plane wavefronts, the coefficient of reflection R and the amplitude transmission coefficient T are given by:

$$R = \frac{V_2\rho_2 - V_1\rho_1}{V_2\rho_2 + V_1\rho_1} \tag{4.22}$$

and

$$T = \frac{2V_1\rho_1}{V_2\rho_2 + V_1\rho_1} \tag{4.23}$$

V and ρ indicate seismic velocity and density of the media, respectively. The term $V\rho$ is the acoustic impedance of the given medium. Clearly, energy will be reflected from the top of a cavern, but poor transmission through the cavity will result in a shadow zone for reflections from horizons below the cavernous one. Cook[7] experimented in salt-mining areas of Texas and the Great Lakes region with conventional seismic equipment. In Texas complex geological structure, and in the Great Lakes experiment strong reverberations, masked any shadow zones which may have existed. Studies were then made on other reflection data of better quality from another unstated salt-mining area. In these studies the ratio of the amplitudes of reflections from an overlying marker horizon and those from the suspected salt horizon were measured. Maxima in this ratio correlated well with positions of brine wells from which salt had been extracted from a horizon some 500 m deep. Borehole-velocity logs were available which confirmed that reflections from the salt horizon had been correctly identified. The use of the amplitude ratio eliminated variations in amplitudes caused by lithological variations in the overburden and differences in geophone coupling. It was also found that this correlation was extracted most readily from records in which automatic gain control had not been used. This last observation is important because generally AGC amplifiers continuously vary their amplification and unless this is monitored it is not possible to reconstruct the original detector (geophone) signal. With the advent of binary gain control equipment, and, of course, full digital recording and data processing, it is possible that seismic reflection could become a powerful method in the search for cavities at considerable depths in areas where the geological structure is not prohibitively complex.

As regards seismic refraction, this method can only be applied indirectly to the problem of detecting cavities in old mine workings and elsewhere by determining variations in the refractor velocity. For this purpose the usual linear shot-geophone array or spread configuration may be advantageously replaced by fan shooting in which shots are placed at the centre of a circular arc and geophones are sited along the arc. This system has been used in mapping salt domes which locally increase seismic velocity, whereas cavities would have the opposite effect. However, it follows from equation 4.20 for the critical distance that the shot-geophone configuration must be carefully fitted to the suspected depth of any old workings. Also, the resultant cavities must be of sufficient size not to lie entirely between two of the radial shot-geophone ray paths. If this all suggests an analogy with a needle in a haystack, then the correct impression will have been created.

A more hopeful, but less-direct approach is to determine the outcrop of faults at rock-head by mapping variations in bedrock velocity. Faults frequently form the limits of underground workings, and any subsidence resulting from the workings probably will be accommodated by movement along it. In this way it may be possible to predict the locus of subsidence at or near ground surface. An example of such a survey is given on page 96.

A further seismic approach to the detection of cavities has been reported by Watkins, Godson and Watson[8] in which the free oscillations induced in cavity walls by seismic charges are detected in standard seismic refraction experiments. Watkins, *et al.*[8] performed their experiments over natural lava tunnels in California and over nuclear explosion cavity *U4B* at the Nevada test site. They also observed amplitude attenuations in the manner described by Cook[7] and delays in arrival times. However, the free oscillations proved the most diagnostic of the three phenomena for the purpose of locating near-surface cavities whose dimensions were as follows:

	Depth to top (m)	Diameter (m)
Lava tunnels	1–8	2–7
Nuclear explosion cavity	25	30

The width of the area in which the free oscillations were observed above the lava tunnels was similar to the surveyed width of the tunnels. in the *U4B* site, the width of the observation area was about twice the width of the cavity. Thus the resolution of the technique depends on the size and depth of the cavity. Cavities are unlikely to be detected either if they are deeper than two or three times their diameter, or if their diameter is less than several geophone spacings (which are generally not less than 1 m each). The depth limitation will depend on the attenuation of waves in the rock overlying the cavity.

Biot's formula can be used to estimate the size of the cavity if the shear-wave velocity in the surrounding material is known; Watkins *et al.*[8] obtained values within a factor of three of the measured sizes.

Case Histories

Magnetometer Survey of Parts of the M1 Motorway

Objective   The object of this survey was to locate old mine shafts along the route of the M1 Motorway across the York, Derby and Nottingham coalfield (Fig. 4.10). The decision to use the magnetic method was made after tests over several known shafts in the neighbourhood of Pinkton and Nuthall in Derbyshire, and after earlier tests using the resistivity method had proved unsuccessful.

Equipment and Field Procedure   The magnetometer used was the Sud Aviation type *MP 102* which is a continuous-reading instrument with an accuracy of 1γ. The field work was carried out by a team of two, one carrying the magnetometer detecting head and the other the recorder. A surveyor was employed on a part-time basis to fix the positions of the magnetic anomalies recorded during the field work.

The area between the Motorway fence lines was surveyed. This area was systematically traversed along lines 3–5 m apart at right angles to the centre line. Whenever a mgnetic anomaly was encountered the particular traverse was stopped and a detailed examination made of the locality to determine the centre of the anomaly and its magnitude. The former was marked by a numbered peg and details of the anomaly recorded in a field notebook. Traversing was then resumed until the next anomaly was encountered.

The rate of traversing was limited only by the walking speed of the two operators and the number of anomalies encountered. The magnetometer was not a limiting factor because of its continuous reading capability.

Results   Subsequent investigation of the anomalies showed that the magnetic method had been successful in detecting:

1. 5 out of 17 mine shafts and bell-pits discovered during the course of construction of the motorway; a further 2 shafts were within 6 m of a magnetic anomaly.

2. 13 roughly hemispherical (in one case pear-shaped) bodies of black shale or ash and burnt shale, 1–2 m in diameter and 1–4·5 m in depth.
3. 11 localities where spoil from old workings or shafts had been dumped.

Other features detected included: a well, 1 m in diameter; 2 septic tanks, 2 m square; an old pond back-filled with rubbish; 5 foundations of old buildings; 5 outcropping coal seams; 11 cases of disturbed ground due to opencast workings; 30 cases of debris left behind after open-cast mining operations. Nearly 20% of the anomalies recorded were associated with land drains, while 13–14% were associated with ground which had not been disturbed in any way.

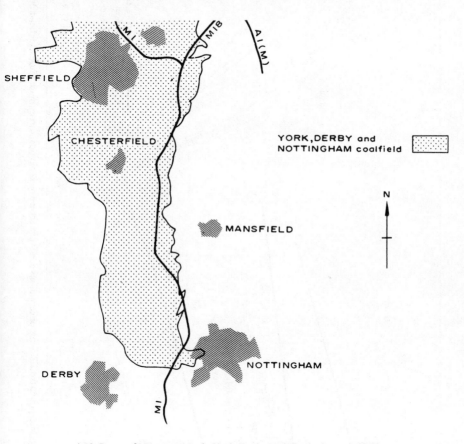

4.10. Route of M1 motorway in York, Derby and Nottingham coalfields

Thus the success rate of the magnetic survey, so far as its primary objective of the detection of old mine shafts and bell pits was concerned, was 30–40%. The other features detected increased the value of the survey from the point of view of site investigation in general. The number of brick structures (foundations of old buildings, septic tanks, etc.) and land drains detected indicates that bricks and drains are frequently slightly magnetic, probably due to the reduction of iron oxides which occurs during the burning of the clay.

96

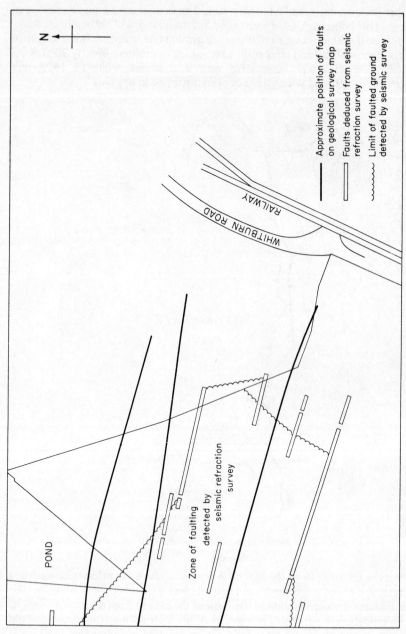

4.11. Faults interpreted from seismic observations and faults shown on geological survey map: Little Boghead development

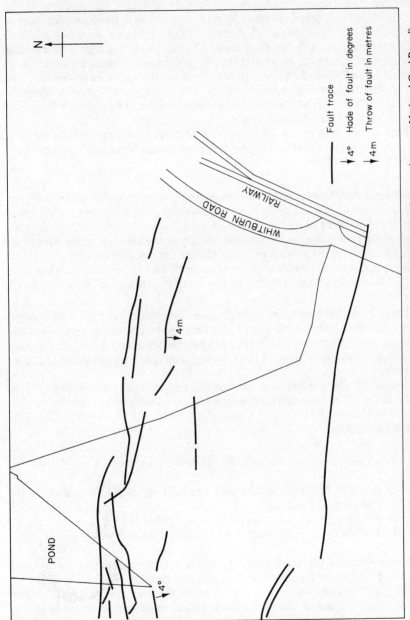

4.12. Faults and other data recorded in underground workings: Little Boghead development (courtesy National Coal Board)

Seismic Refraction Survey for the Little Boghead Development

Objective The objective of this survey was to locate the position of the Heatherfield Fault below the site (Fig. 4.11). According to the Geological Survey of Scotland Memoir on the area[9], the fault had two branches, a northerly branch with a downthrow to the south of about 25 m and a main branch with a similar downthrow estimated at about 100–110 m. 'A deep boring put down about 660 m west of Little Boghead (Vivians Bore) . . . passes through a large fault . . . The throw of the fault cannot be less than 92 m.' The 6 in to 1 mile Geological Survey of Scotland, Sheet *IXNW*, Linlithgowshire, published in 1898, shows a third fault branch between the other two (Fig. 4.11).

Bedrock was covered to a depth of between 3 and 12 m by superficial deposits, consisting of recent alluvium and peat over boulder clay containing layers or lenses of sand and gravel.

Equipment and Field Procedure A preliminary test of the resistivity and seismic refraction methods was carried out. The resistivity measurements gave no indication of the presence of faulting in the bedrock, but the seismic method recorded marked lateral changes in the seismic velocity of the bedrock which were interpreted as marking the position of faults. The test showed the existence of two main faults crossing the site which were correlated with the northern and main branches of the Heatherfield Fault. The survey was completed, therefore, using the seismic method.

Two 12-channel seismic equipments, coupled together to form a 24-channel unit, were employed. Seismic energy was obtained by detonating small explosive charges, varying between 100 and 300 g in weight, buried about 1 m below ground level. A geophone separation of 10 m, corresponding to a spread length of 220 m, was used.

A total of thirteen seismic spreads, oriented approximately at right angles to the strike of the faults were carried out by a team consisting of a geophysicist, electronics technician and surveyor. A geologist assisted with the interpretation of the field measurements.

Results The bedrock seismic velocity values recorded were:

1. South of the main branch and north of the northerly branch of the Heatherfield Fault: 2625–3215 m/s.
2. Between the two branches of the faults: 2065–3725 m/s.
3. The seismic velocity range of the superficial deposits was 492–1245 m/s.

The positions of the lateral bedrock seismic velocity changes recorded during the survey indicated a zone of faulting crossing the site, as shown in Fig. 4.11, rather than the three distinct faults shown on the Geological Survey map. This interpretation was supported by the fault pattern recorded in the underground workings of the Wilsontown Main Coal (Fig. 4.12). On the western side of the site the fault zone indicated by the seismic survey extended 25–30 m north of the assumed position of the northern branch of the Heatherfield Fault shown on the Geological Survey map. On the eastern side of the site, faulting was shown to extend 50–55 m south of the assumed position of the main fault branch.

Resistivity Survey at the Woodcross Lane Clinic Site, Wolverhampton

Objective   Two old mine shafts were known to occur on the site and the objective of the resistivity survey was to determine whether other old shafts or sub-surface voids were present within 9 m of the ground surface.

The resistivity method rather than the magnetic method was used for the survey, because an overhead power line and underground services near to the site would have interfered with the latter.

Equipment and Field Procedure   Resistivity measurements were made, using standard Terrameter equipment, at a grid of points 3 m apart, covering the whole site. Two resistivity measurements were made at each point using different electrode arrays called dipole-dipole and pole-dipole, illustrated below. The former gave a depth of penetration of about 3 m while the latter gave a depth of penetration of up to 10 m.

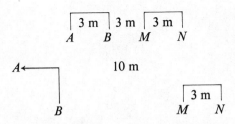

Iso-resistivity maps were prepared from the measurements obtained with each of the electrode arrays (Figs. 4.13 and 4.14).

Results   The two known mine shafts were closely associated with areas of low resistivity for both the electrode arrays. A recommendation was made, therefore, for other low resistivity areas on the site to be investigated by trenching.

A sub-surface void, whether air filled or water filled, would be expected to be associated with high resistivity. On the Woodcross Lane site, however, the high resistivity areas were strongly marked on the dipole–dipole isoresistivity map but were relatively insignificant on the pole–dipole isoresistivity map. Therefore it was concluded that the high resistivities were associated with surface rubble rather than sub-surface voids.

Conclusions

Each of the geophysical methods previously described presents a problem of scale in the detection of old mineral workings. Geophysical measurements on the ground surface have to be spaced at distances such that anomalous fields can be detected, and the shape and extent of these fields is dependent on the size and depth of the anomalous body. The subsurface materials between the anomalous body and the ground surface behave as a filter which produces a window; if the observations are not aimed through this window, they will not detect the anomalous field and hence the target will be missed. It is therefore necessary to use all available information to estimate the approximate location, depth and size of old workings, as

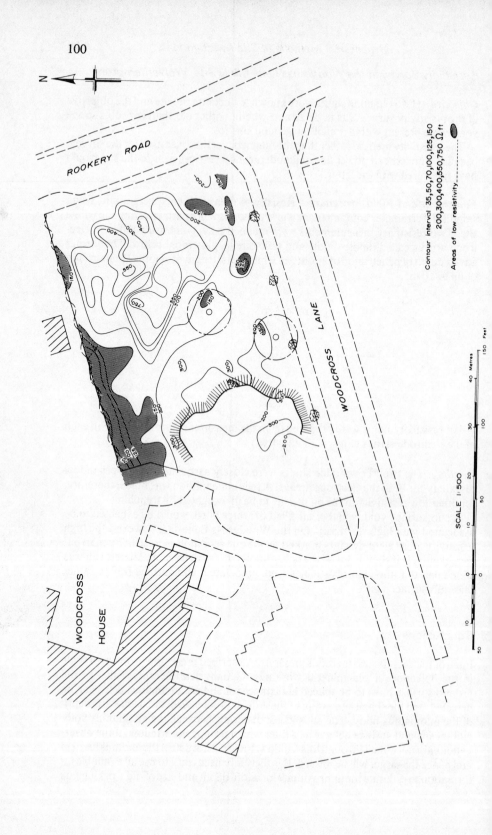

N

ROOKERY ROAD

WOODCROSS HOUSE

WOODCROSS LANE

Contour interval 35,50,70,100,125,150
200,300,400,550,750 Ω ft

Areas of low resistivity ----------

SCALE 1:500

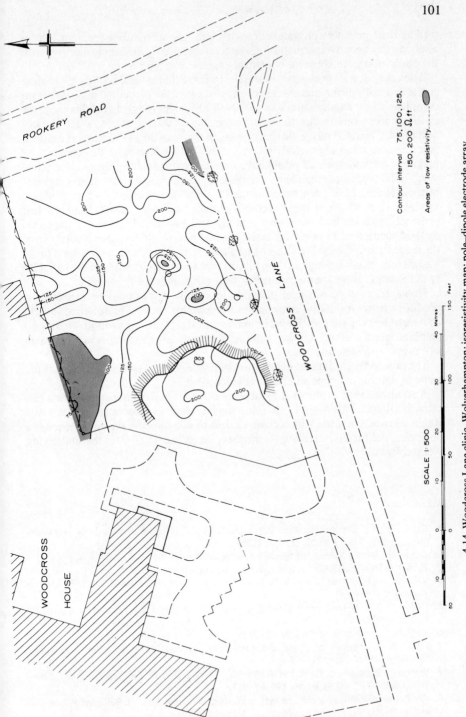

4.14. Woodcross Lane clinic, Wolverhampton: isoresistivity map; pole–dipole electrode array

well as their probable physical contrasts with the surrounding medium, before applying any geophysical method. Even then extraneous fields may interfere with the observations, or some other contrary factor may intervene.

The case histories previously described indicate that in the detection of old mine shafts and bell-pits the success rate of any one geophysical method may vary from complete failure on some sites to a possible 100% success on others. This variation in the success rate is due to variations in the character of the target already described. Clearly, if all the old shafts along the route of the M1 Motorway had had precisely the same characteristics as the 30–40% which were detected by the magnetic method, then all would have been discovered during the magnetic survey. The fact that more than half were missed, however, indicates that the magnetic method should have been supplemented by one or more of the other methods available. But which of these methods should have been used? The electrical resistivity method had been tried already and had failed where the magnetic method succeeded. Possibly the resistivity method would have succeeded where the magnetic method failed. This example illustrates the dilemma in which those who attempt to advise engineers on this problem are placed; owing to the variability of the target there are no certainties, only possibilities.

However, two geophysical methods in combination could be expected to produce a better result than one method alone simply because two methods would cover a wider range of possible variations in the target than one method. There is no combination of methods known to the authors at the present time, however, which will ensure complete success.

The position regarding the detection of old underground workings is broadly the same as that for old mine shafts already described.

A solution to the problem of locating fault zones in Coal Measures strata can often be found with the seismic refraction method as illustrated by the Little Boghead case. On a site with a cover of drift or alluvium the seismic method is a powerful tool for investigating the quality, including faulting, of the underlying worked strata.

REFERENCES

1. Taylor, R. K., 'Site Investigations in Coalfields—the Problem of Shallow Mine Workings,' *Quart. J. Eng. Geol.,* **1** No. 2, 115–133 (1968)
2. Briggs, H., *Mining Subsidence,* Edward Arnold, London (1929)
3. *Report on Mining Subsidence,* Inst. Civ. Eng., London (1959)
4. Van Nostrand, R. G. and Cook, K. L., *Interpretation of Resistivity Data,* U.S.G.S. Professional Paper No. 499 (1966)
5. Colley, G. C., 'The Detection of Caves by Gravity Measurements,' *Geophysical Prospecting,* **11** No. 1, 1–9 (1963)
6. Nagata, T., *Rock Magnetism,* Maruzen Press, Tokyo, 2nd edn (1961)
7. Cook, J. C., 'Seismic Mapping of Underground Cavities Using Reflection Amplitudes', *Geophysics,* **30** No. 4, 527 (1965)
8. Watkins, J. S., Godson, R. H. and Watson, K., *Seismic Detection of Near Surface Cavities,* U.S.G.S. Professional Paper No. 599-A (1967)
9. *The Economic Geology of the Central Coalfields of Scotland,* Area VI, Bathgate, Wilsontown and Shotts, Geological Survey of Scotland Memoir (1923)

Chapter 5

Improved Geophysical Techniques for Survey of Disturbed Ground

Two new techniques have been developed which appear to provide improved geophysical methods for locating a variety of sub-surface features, particularly old mine workings, filled quarries, cavities, faults, culverts and similar phenomena. The two methods are described below.

Current-path Electromagnetic Prospection

This method requires the application of an oscillating electric current to the ground by means of two electrodes spaced from 100 to 500 m apart. The current source is an earth current generator (ECG). This is a small battery operated unit generating a high impedance output and with the means for controlling the current flow into the ground. Light cable connects the ECG output to the electrodes. The latter are pushed into the ground to a depth of about 400 mm and the cable is run well clear of the line between them.

Where the ground is homogeneous the current flow may be regarded as flowing in substantially undeviating paths (filaments) over the central portion of a traverse between the two electrodes. However, where a cavity or filled feature exists the current paths deviate and the essence of the method is to locate the points of deviation.

The equipment used to detect these deviations is shown in Fig. 5.1 and consists essentially of a theodolite-mounted aerial array which can measure the direction of the resultant oscillating magnetic field of the ground currents. This may be conveniently read in azimuth and in elevation co-planar with the traverse. The unit is mounted on a tripod and incorporates a small receiver which is sharply tuned to the frequency of the ECG. The aerial is swung successively in the two planes until a minimum or null signal is obtained on a head-set worn by the operator.

The procedure in the field is to set out an initial traverse between the electrode points, and proceed along this taking observations at intervals, typically 5 or 10 m apart. Under average conditions, around sixty observations can be taken in a day, though with practice this rate may be considerably exceeded. Hence between 300 and 600 m of traverse length can be executed in a day on site. The site to be surveyed is covered by a requisite number of such traverses, the traverse separation being selected according to the purpose of the survey. Traverse separations usually exceed 5 m, and may be hundreds of metres apart where a search is being made for large features.

5.1. Current-path electromagnetic prospection equipment in use in the field

With the completion of the observations, the results are profiled on graph paper (Fig. 5.2). This shows the aerial alignments found at each observation point.

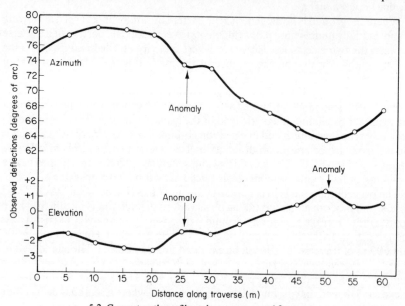

5.2. Current-path profiles taken over ground features

Ground anomalies are revealed by local deviations of the readings and they suggest the positions where sub-surface discontinuities may be suspected. Confirmation of features should be made by excavation or drilling.

The example given in Fig. 5.2 is of a profile taken along a traverse over what proved to be a swallow hole in limestone under the route of a motorway. It was found on excavation to be about 3 m in diameter and its roof lay 2·5 m below the traverse line. The profiles show deviations from the mean line in both elevation and azimuth.

There is obviously a relationship between the depth at which a particular feature can be located, and its physical size. The method has been found suitable to locate features where the ratio of depth : diameter is of the order of 8 : 1 or smaller. Although the depth of the feature cannot be quantitatively determined, some assessment regarding its proximity to the surface is possible.

On account of interference from metal pipes and cables buried in the ground, the current-path method is most satisfactorily employed on open sites.

Seismic Reflection

The use of seismic reflection techniques for deep strata prospection during petroleum exploration is well established. However, the use of this method for probing shallow depths which are of interest to civil engineers has not been investigated until quite recently. With this new seismic reflection method it is possible to assemble on paper a representation of the reflected ground motions occurring as a result of a series of rammer-induced shocks.

The procedure requires field equipment (Fig. 5.3) consisting of a 6 kg rammer

5.3. Seismic-reflection field equipment

Improved Geophysical Techniques

with an attached transducer and proximity switch, a twin-channel tape recorder, a belt-slung control pack and a geophone.

The rammer is dropped to the ground at a distance of about 0·5 m from the geophone. At a rammer height of 150 mm, a sliding-rod proximity device operates a pulse generator which causes the tape recorder to run for 500 ms. At the moment of rammer impact the transducer causes a datum pulse to be applied to one channel of the tape recorder, while the other channel records the displacements of the main geophone diaphragm caused by the initial and reflected ground motions. This takes place over a period of approximately 450 ms. This sequence is repeated at intervals of from 100 mm to 5 m normally along a straight traverse line. The sounding separation distance depends on the purpose of the survey. About 400 impacts per hour are easily possible on a surface clear of major obstructions.

A small, local feature, which gives rise to detectable reflections, will cause reflection events to appear only on a few adjacent soundings of, say, a 100-sounding series. Conversely, a large-scale plane feature, e.g. bed-rock, will cause reflection events to appear on the reflection record of a large number of sounding positions.

A simple method of achieving a clear determination of which sounding positions are subject to local reflections as well as general reflections, and which positions lack local reflectors, is to assemble the sounding reflection records side by side on a

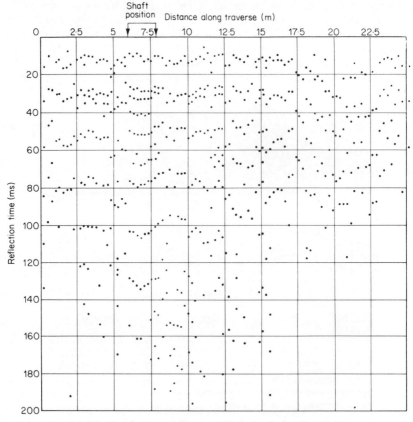

5.4. Seismic-reflection profile over a filled mineshaft at St. Helens

sheet of paper. This is done by using a digital plotter, which, after re-recording and filtering of the field tape, prints on a chart successive sets of arrivals received at each sounding point along the traverse. The resulting dot-density pattern, showing the reflection events under each traverse, can then be interpreted to yield useful information about the presence of sub-surface features. Fig. 5.4 gives an example of a chart profile. This is taken from a traverse across an old mineshaft which was filled with waste material to its depth of 60 m and concealed under a layer of 4·5 m of fill material. The formation of a distinctive reflection pattern, indicative of the presence of a filled shaft, can be seen at the point marked. This was subsequently confirmed by drilling.

Like other seismic methods, this technique provides a depth profile. In fact, it is possible to select the vertical scale at which the charts are printed, and this can prove useful in the correlation of boreholes. As an illustration, a good reflecting interface at a depth of 50 m may be expected to cause a reflection to occur at about 160 ms after impact. At a depth of 100 m the reflection time would probably be slightly less than twice as long, e.g. perhaps 280 ms (the average vertical velocity of the upper medium tending to increase as the depth increases). For such features a 400 ms vertical scale would be appropriate. A cavity at a depth of 2 m, on the other hand, would occasion a reflection at less than 15 ms. Such a feature would be

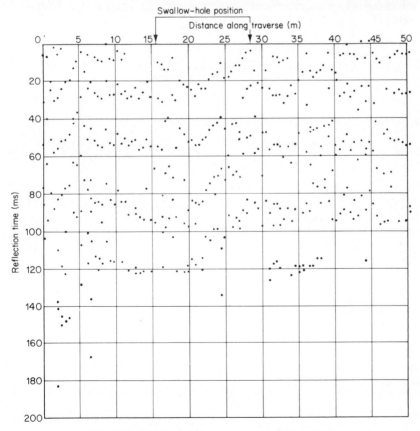

5.5. Seismic-reflection profile over a swallow-hole at Radyr

displayed in more vertical detail on a print-out with a 50 ms vertical scale.

The interpretation of the reflection profiles is assisted by the fact that they often reflect the shape of the section of the feature they are investigating, e.g. the profile over a swallow-hole (Fig. 5.5) where the 'V'-section of the mouth causes successively later and then earlier reflected arrivals as it is traversed.

The profile across the mineshaft (Fig. 5.4) emphasises another important characteristic of seismic reflection, i.e. its lateral discrimination. The reflection events due to the shaft cease abruptly on either side, even though it is only about 2 m wide and buried below nearly 5 m of fill. This implies that the system responds only to near vertically ascending wavefronts.

The equipment can also be used for seismic refraction work. In this case hammer blows are all made at the same sounding point, the geophone being moved at intervals of 0·5 or 1 m until about twenty observations are made. In the print-out seen in Fig. 5.6 the first arrival signals align themselves as a diagonal row of dots on the chart. From these it is possible to draw a slope giving the average vertical seismic velocity under the short traverse taken and derive the depth to the first refracting interface. The process can be repeated by drawing a diagonal line related to the average horizontal seismic velocity of the material underlying the overburden. An interesting feature of this technique is that it allows seismic-refraction soundings to be taken relating to the material under a hard top surface such as concrete or tarmac.

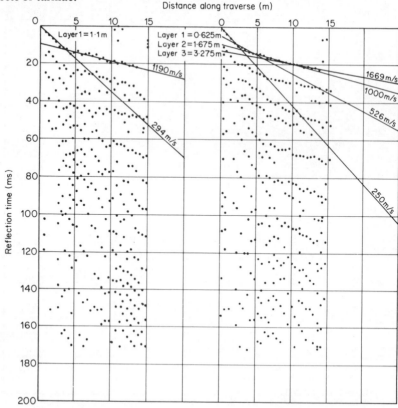

5.6. Seismic-refraction profile to determine seismic velocities of different layers at Taunton

Chapter 6

Mining Subsidence

Coal has been worked in the UK for approximately a thousand years, starting with open-casting and shallow working of outcrops and developing to deeper and more extensive extractions with time and technological advancement. The vast majority of the coal mined in the UK at the present time is by panel working, a development of the longwall system, which is suited to mechanised extraction techniques. In this system the roof in the area of extraction behind the working face (i.e. in the goaf) is held up for a short distance by temporary supports and beyond these the zone of roof break-up commences. The part in the latter zone nearest the face is poorly compacted and voids are formed above the seam. Further still from the face, the workings are subjected to increasing overburden loads and so the roof and floor converge and the voids close. Consolidation continues until it is in equilibrium with the strata load. Disturbance becomes progressively reduced above the extracted coal seam. It is usually manifested by a gentle sagging of the strata or movements along bedding planes and joints. The final observable surface movements occur shortly after the working face has passed beneath the site.

The early methods of working by bell-pits and room and pillar systems together with their associated problems are described in Chapter 7. Since the majority of present-day problems result from recent and current methods of working, the following remarks apply generally to total extraction.

History and Early Theories

Subsidence at the surface and damage to buildings resulting from the underground working of coal was recognised by the beginning of the nineteenth century. For example, damage in the city of Liège, Belgium, resulted in the appointment of commissions in 1825 and 1839 which concluded that workings deeper than 100 m would not affect the surface. At this time the *vertical* theory was generally accepted, according to which subsidence was considered to occur vertically above the mined area only (Fig. 6.1a). Consequently it was assumed that leaving a pillar of coal, of similar shape and size, vertically below a building would be sufficient to protect it and to prevent damage. However, it was soon observed that damage occurred to structures beyond the 'vertical' limits and this theory was therefore superseded by the *normal* theory of Gonot, a Belgian engineer whose work was published in 1858. This theory proposed that subsidence occurred within an area defined by the normals projected from the working faces to the ground surface

(Fig. 6.1*b*). The consideration of this theory together with associated research was started by both Gonot and a French engineer called Toillez as early as 1838.

The normal theory was criticised in 1867 by Schulz and Sparre, two German engineers, who separately concluded that with dipping beds of shale the influence

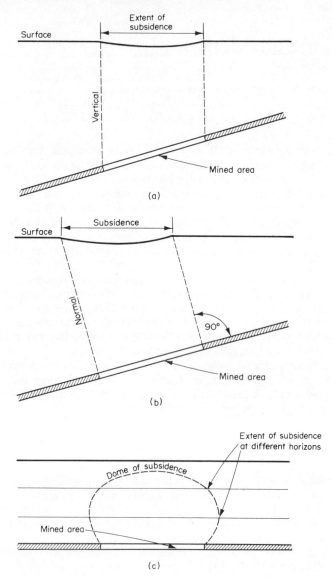

6.1. Subsidence theories. (*a*) Vertical, (*b*) normal and (*c*) dome

would be vertical, while in sandstone it would be vertical on the rise side. Criticism of the theory was also offered by Dumont (another Belgian engineer) in 1871, whose observations at Liège led him to claim that the normal theory could not be applied to workings steeper than 68°.

In 1885 the famous Frenchman Fayol reviewed the past work and made numerous observations both in the field and on laboratory models from which his *dome* theory emerged (Fig. 6.1c). He suggested that in horizontally bedded strata the movements of the ground were delimited by a kind of dome which was approximately elliptical in shape (in dipping rocks the axis of the dome was inclined at right angles to the beds). He proposed that as the width of the workings increased the height of the dome also increased so that there would be for any given width of working a 'safe' depth from which the dome would not extend to the surface.

Halbaum[1] considered the strata above the goaf to act as a cantilever beam. He assumed that its neutral axis was near its base due to the greater strength of rock in compression than in tension. After conducting tests on a series of rock types present in the Coal Measures he concluded that up to the neutral axis the line of fracture of the strata would hade back over the goaf, whilst above this axis the hade would be reversed over the solid rock. In this manner he was able to explain the 'angle of influence' whereby the subsidence profile extends over a greater area on the surface than that immediately above the goaf. Halbaum's cantilever theory also provides an indication of the surface strains likely to be generated by subsidence. The character of the stresses in a beam reflect its curvature in that they are tensile on the convex side and compressive on the concave side. Moreover, the magnitude of the surface strains produced by subsidence is inversely proportional to the radius of curvature of the surface on which the strains are measured.

In 1929 Briggs[2] published his classical work entitled *Mining Subsidence* in which he wrote 'mining subsidence is not amenable to mathematical analysis . . . empirical formulae are sometimes useful as guides and will probably become more so as experience extends and is made more available', and 'measures are able considerably to reduce the chances of damage . . ., but though we can lessen the risk and rigours of injury we shall never altogether remove that element of uncertainty which is to be regarded as indigenous to mining subsidence'. These were fair comments at the time. However, it is now generally accepted that the many detailed observations of surface ground movements, together with the analyses of the data collected, enable the general pattern of the movements to be defined. Hence ground movement can now be accurately predicted and graphical methods based on modern research are embodied in the *Subsidence Engineer's Handbook*[3].

The Ground Movement Process

The logical pursuit of ground movements from the underground excavation through the overlying strata to the surface and the resultant damage to structures thereon can be resolved into three processes as follows.

1. *Mining factor(s)*. The basic or orthodox movements which develop at the interface (top of the Coal Measures) as a result of underground extraction. These are mathematically predictable and are governed by the principles of rock mechanics and vary with mining dimensions of section extracted, depth from surface, extent of workings, etc.
2. *Site factor(s)*. The variation or measure of behaviour of the site due to the nature of rocks and soils when influenced by discontinuities such as jointing and faulting, changes in hydrogeology, and anthropogenic modifications to the surface including residual effects of previous mineral extraction.

3. *Structural factor(s).* The tolerance of surface structures towards the transmission of effective ground movement, influenced by: shape, size, design of foundation and super-structure, materials used, age, standard of maintenance and repair, and purpose for which the structure is or is to be used.

This method of approach and the relationship of the three main process factors is illustrated in Fig. 6.2. There are many permutations of the conditions comprising each factor, with combinations of all three. However, for the sake of convenience they are dealt with separately as follows.

BASIC GROUND MOVEMENTS (MINING FACTOR)

When coal is extracted in sufficient quantity underground, the roof beds eventually subside into the excavated void and movements are transmitted through the overlying strata to the surface. The movements which develop at the surface are:

1. Vertical subsidence.
2. Tilt: differential subsidence.
3. Curvature: differential tilt.
4. Horizontal displacement.
5. Strain, extension (+) and compression (−) differential displacement.

The relationships of these parameters of ground movement to the underground excavation are shown diagrammatically in Fig. 6.3. From this diagram it will be observed that the extent of affected surface is greater than the worked area in the seam. The limit of the effect on the surface is defined by the 'angle of influence' which has been determined by precise observations in British coalfields to average 35° to the normal of the seam. In a level seam the greatest amount of subsidence occurs over the centre of the working, diminishing to zero approximately at 0·7 of the depth outside the boundaries of the excavation. However, for practical purposes of noticeable movement and damage, a distance equal to one half of the depth is more appropriate.

The concept of the 'area of influence' or critical area of extraction originated in Germany and was further explained by Wardell[4]. For a given point P on the surface this area in a level seam is the circle at the base of an imaginary cone with its axis passing through the strata and P at its apex (Fig. 6.4). The diameter of the area is equal to 1·4 times the depth of the seam. Workings outside the area cannot affect point P whereas all workings within the area must have an effect on P. It follows that all the coal from within the area must be extracted before the point can undergo maximum subsidence. If only part of the critical area is worked then point P can only suffer partial subsidence.

In relation to the critical area, a given working, depending on its width, will be (a) sub-critical (width less than 1·4 times the depth), (b) critical (width equal to 1·4 times the depth) or (c) super-critical (width greater than 1·4 times the depth). The maximum possible subsidence is only achieved when the width of working is either critical or greater than critical (Fig. 6.5). If the extracted void is left untreated the maximum subsidence is normally 84–90% of the extracted thickness; if the void is solidly stowed, subsidence is approximately halved. Orchard[5] has shown that the amount of subsidence for a given extraction can be related to the width : depth ratio of the working as shown in Fig. 6.6.

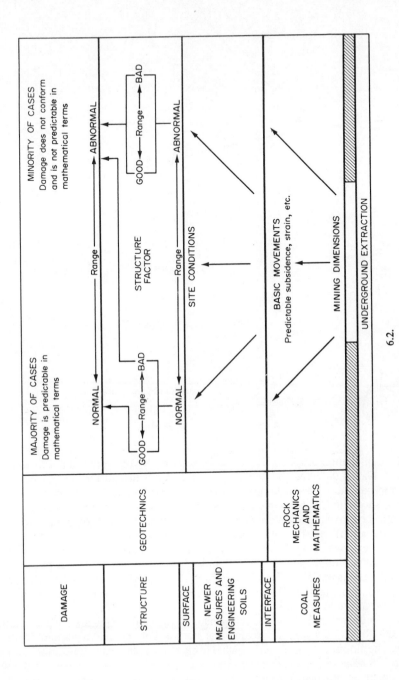

6.2.

Mining Subsidence

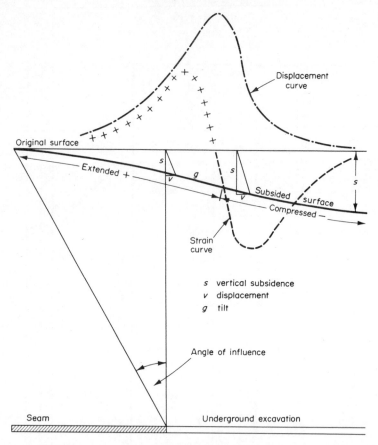

6.3. Elements of ground movement

Trough-shaped subsidence profiles develop tilt between adjacent points which have subsided different amounts. On an area of affected surface points subside downwards and are displaced horizontally inwards towards the axis of the excavation as shown in Fig. 6.7. Differential horizontal displacements result in a zone of apparent extension on the convex part of the subsidence profile (over the edges of the excavation) whilst a zone of compressions develops on the concave section over the excavation itself. Simplified zones of extension and compression are shown diagrammatically in Fig. 6.7 and strain values are usually expressed in parts per thousand or millimetres per metre. In addition to subsidence, Orchard has also shown that the values for tilt and strain parameters are also related to the width : depth ratio (Fig. 6.8).

Maximum ground tilts are developed about the limits of the area of extraction and may be cumulative if more than one seam is worked up to a common boundary. The amount of movement which is likely to take place can be assessed before the coal is mined and this can be allowed for in the design, assuming that extraction will occur as planned. Perhaps, after discussion with the mine operators, rephasing or layout changes may be put in hand so that the additive effect of working

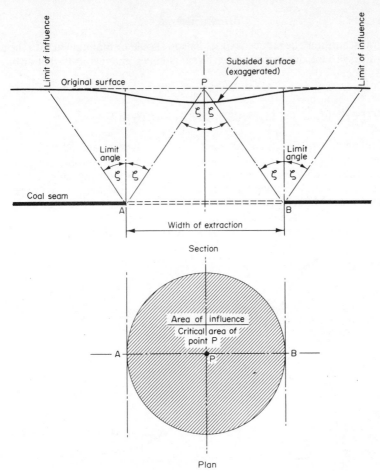

6.4. Area of influence

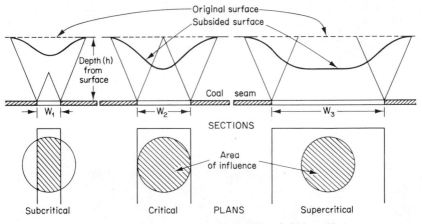

6.5. Change in surface subsidence by variation of width working

several seams may be reduced. Close liaison should be maintained with the mine operators in case, due to unforeseen circumstances, changes in extraction have to take place.

GEOTECHNICAL INFLUENCES (SITE FACTOR)

The magnitude of the forces imposed on the rock mass are determined by the dimensions of mineral extraction; likewise, the movements which are induced at the surface are influenced by variations in the site condition, especially by the near-surface rocks and engineering soils together with changes in hydrogeology. Furthermore, Mather, Gray and Jenkins[6] in their investigation of ground-water occurrence at Aberfan concluded that hydrogeology can be affected by mining subsidence. The general mechanical properties of the near-surface rocks in the

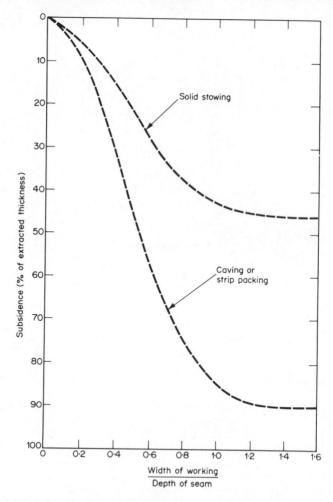

6.6. Relationship of subsidence to the width : depth of working (after Orchard[5])

coalfields vary from brittle sandstones and limestones through transitional siltstones to plastic mudstones and shales.

Fissuring may result from stretching on the convex surface of the subsidence basin. The necessary readjustment in weak strata can normally be accommodated by small movements along joints. However, as the strength of the surface rock and the joint spacing increases so the movement tends to become concentrated at fewer points. For instance, in massive limestones and sandstones movement may be restricted to the master joints. At rock-head this can cause such joints to gape up to 0·5 m.

Sandstones often react in an orthodox manner, with small joints opening and closing in areas affected by extension and compression, respectively.

Bunter sandstone sometimes fissures when affected by differential displacement, especially in areas of ridge and valley configuration where jointing patterns are prominent. The massive sandstones of the Coal Measures have also suffered

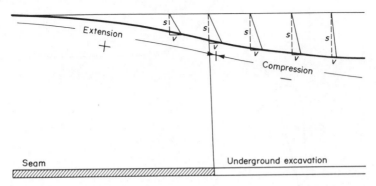

Section on *A–B*

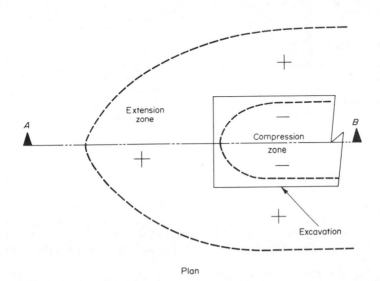

Plan

6.7. Development of strain zones. Key as for Fig. 6.3.

much fissuring as a consequence of subsidence. The exact location and magnitude of these movements cannot be easily predicted and therefore potential lines of fissuring cannot be treated on an individual basis.

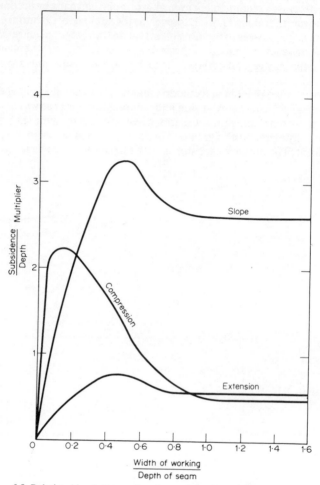

6.8. Relationship of slope and strain to the width : depth of working

The Magnesian Limestone of the Permian system is brittle and variably bedded with well-developed systems of jointing related to regional tectonics, as has been shown by Moseley and Ahmed[7]. However, the prominence of joints is greatly influenced by the extent to which the host rock has been weathered, and cambering in Permian limestone, which is associated with valley bulging, has tended to open joints thus forming 'gulls' or 'V'-mouthed fissures. This type of site condition is likely to react in an unorthodox manner when affected by mining, differential displacement (strain) often concentrating at particularly well-developed joints or fissures.

In clay deposits the behaviour is near isotropic, except in the case of gypsiferous marls containing solution cavities.

Faulting has, in a large number of cases, contributed to major subsidence damage. This is because surface cropping faults may react in an abnormal manner when affected by underground extraction. Abnormal movement may be either a concentration of horizontal strain or a vertical stepping of the ground on one side of the fault. The degree of probability of such an occurrence and the magnitude of stepping are influenced by the mining dimensions and the extent of working. Nevertheless, there is no reliable method for obtaining the degree of differential movement that may occur across a fault, but experience has shown that movements are frequently greatest where the workings are under the hade of the fault and where a single shear plane rather than a shatter zone is present. Where faults have been met with during previous phases of mining operations their surface outcrop can frequently be projected. Furthermore, their outcrop may display itself in a zone of marked surface movement or damage. If this is not so, then field mapping, trenching or drilling will be required to determine their positions.

STRUCTURES (STRUCTURE FACTOR)

It is not possible to allocate the equivalent of a numerical factor to the different types of surface structure which can be affected by underground mining. Such an assessment is largely based on experience and an understanding of the inherent strengths and weaknesses of that particular structure, and how it will react when affected by movement. The various factors which must be taken into account are:

1. Size and shape.
2. Design of foundations.
3. Type of superstructure.
4. Methods of construction and quality of materials.
5. Age and standard of maintenance and repair.
6. Purpose for which the structure is used, etc.

For example, an old dilapidated building suffering from the effects of lack of maintenance and repair, will be more sensitive and will react more violently than a similar building which has been well maintained and properly repaired, when both are subjected to the same amounts of movement.

Subsidence Damage

The assessment of effective ground movement is usually required in order to consider the effect of particular workings on a given structure or structures. It is obvious that the individual elements of ground movement have different effects and varying roles of importance for different types of structure. For instance, vertical subsidence is the most important type of movement in low-lying areas which are subject to flooding and drainage problems. A few centimetres of subsidence can sometimes result in very costly damage. Tilt also causes concern. It may damage drainage works, it presents problems with communication structures such as highways, canals and rail tracks, with tall structures like chimneys, and it brings about the dislevelment of industrial machinery.

Damage to buildings and conventional structures is generally caused by differential horizontal movements and the concavity and convexity of the subsided profile resulting in compression and extension in the structure itself. Many materials and structures are stronger in (and are more capable of resisting) compression than extension. Most buildings require approximately twice the amount of compression to develop damage comparable to that caused by a given amount of extension.

Typical mining damage starts to appear in conventional structures when they are subjected to effective strains of 0·5–1·0 mm/m. Damage is generally classified by the scale (see Table 6.1):

1. Very slight or negligible.
2. Slight.
3. Appreciable.
4. Severe.
5. Very severe.

This was introduced by Orchard some years ago together with the damage classification graph (Fig. 6.9). It is emphasised that it is intended that the graph and scale are only to be used by engineers possessing the necessary skill and expertise, when it can then prove to be a valuable approach to prediction and assessment.

Pseudo-mining Damage

The occurrence of typical mining damage to structures which have not been affected by mining or which are remote from mining areas is one which is often reported. Natural phenomena which cause movement of the ground can result in damage which is similar in appearance to that occasioned by mining. Clay shrinkage and expansion causes damage which may be confused with that from mining. The same applies to absorption of water from soils by vegetation—poplar trees are often responsible for appreciable localised settlement. Similar effects occur when sub-soils are scoured away from foundations by water flowing from faulty drains. The collapse of a drain running beneath or close to a wall may result in a corresponding collapse in the brickwork above.

Subsidence may occur when the natural water level of the locality is lowered by pumping, as in south east Durham, where some surface movement and damage is attributed to the de-watering of the Magnesian Limestone. Subsidence of more than 6 m has occurred in Mexico City over the last 60 years due to water abstraction. Similar movements have been experienced in Houston, Texas, and in the Santa Clara valley in California. More locally, rises and falls of up to 100 mm have been observed in Nottinghamshire on the Lower Mottled Sandstone and have been related to seasonal changes in precipitation and ground-water levels.

Weak foundations, building malpractices and the use of unseasoned timber, etc. may give rise to fractures and damage in the fabric of buildings which, to all intents and purposes, are similar to those caused by mining. Other causes of damage are those from sulphate attack, rusting of ferrous metals, thermal movements and roof spread.

It is most important that these forms and causes of pseudo-mining damage be recognised, since many claims for such damage are made against mineral operators and have, of necessity, to be repudiated as being more properly attributable to one or more of the other causes just mentioned. In some cases there may be a mixture of both mining and other damage, and the total claim may have to be assessed and resolved accordingly.

Table 6.1 NATIONAL COAL BOARD CLASSIFICATION OF SUBSIDENCE DAMAGE

Change in length of structure (mm)	Class of damage	Description of typical damage
Up to 30	1. Very slight or negligible	Hair cracks in plaster. Perhaps isolated slight fracture in the building, not visible on outside
30–60	2. Slight	Several slight fractures showing inside the building. Doors and windows may stick slightly. Repairs to decoration probably necessary
60–120	3. Appreciable	Slight fracture showing on outside of building (or main fracture). Doors and windows sticking, service pipes may fracture
120–180	4. Severe	Service pipes disrupted. Open fractures requiring rebonding and allowing weather into the structure. Window and door frames distorted; floors sloping noticeably; walls leaning or bulging noticeably. Some loss of bearing in beams. If compressive damage, overlapping of roof joints and lifting of brickwork with open horizontal fractures
> 180	5. Very severe	As above, but worse, and requiring partial or complete rebuilding. Roof and floor beams lose bearing and need shoring up. Windows broken with distortion. Severe slopes on floors. If compressive damage, severe buckling and bulging of the roof and walls

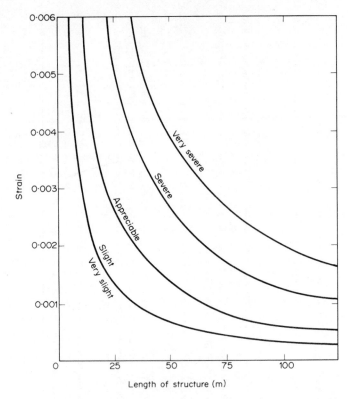

6.9. Relationship of damage to length of structure and horizontal strain (after Orchard[5])

The Control of Mining Damage

The effects of mining subsidence at the surface and the resultant damage are capable of assessment. The damage to conventional structures can be controlled and influenced by:

1. Precautionary measures built into new structures in mining areas.
2. Preventive works applied to existing structures that are to be affected by underground mining.
3. Mine design incorporating special underground layouts.
4. Any combination of 1–3.

PRECAUTIONARY MEASURES

The most common method of mitigating subsidence damage is by the introduction of flexibility in design; various methods are catalogued in the N.C.B. Subsidence Engineer's Handbook[3] including: the use of reinforced-concrete rafts laid on granular material to reduce the friction between the ground and the structure;

planned discontinuities in large structures; and arrangements for controlled distortion as used in the C.L.A.S.P. type of buildings and the storage reservoirs of the Central Nottinghamshire Water Board. Flexible joints in services, jacking devices for correction of tilt in structures and machines, flexible pavements and rocking bridges in highway construction are other examples which benefit from flexibility in design.

Methods and materials aimed at flexibility do not necessarily mean higher costs, in fact they often have the advantages of reduced costs, especially in the cost of maintenance after construction and on naturally unstable sites. Emphasis is placed on flexibility because in the vast majority of cases it is either too costly or it is physically impossible to design rigid structures which are capable of resisting mining movements.

PREVENTIVE WORKS

Preventive works can often be used to reduce the effect of movements on existing structures. Again, one of the main aims is to achieve greater flexibility. For instance, some buildings can be severed to reduce large structures to smaller units. The cutting of buildings can sometimes be carried out at connecting corridors, the cuts being made weatherproof with flexible coverings.

In areas which are to be subjected to appreciable compressive strains, 'trenching' around large buildings down to below foundation level and backfilling with compressible material will often result in a reduction of effective compression and associated damage by about 50%.

If structures which are weak in tensile strength are likely to be subjected to appreciable extension, they can be given some measure of protection by 'strapping' or tie-bolting together.

The supporting of arches, shoring of walls and insertion of flexible joints in services are often carried out. With all cases of preventive works, the expense should be justified unless there are circumstances which make it imperative to minimise damage regardless of cost as in the case of hospitals and schools, etc.

MINE DESIGN

Mine design usually takes account of the surface and is often influenced by surface developments. Underground workings can be projected so that structures which require to be considered are subjected to movements which are likely to result in minimum damage. Cost is taken into account in these cases and obviously economic extraction sometimes requires a compromise.

Special layouts are sometimes necessary. Harmonic extraction has been used on the continent where simultaneous multi-seam working can be practised. Harmonic extraction is a method of working which has been devised to reduce the strain associated with subsidence rather than its actual amount by working adjacent panels in different seams so that the tensile stresses induced by one face are partially counteracted by the compressive stresses induced by the second face.

Stepped faces have been used successfully in Britain. With this method, two adjacent faces separated by a time lag are worked in the same seam. The tempory travelling stresses of the two panels tend to counteract each other. The principle involved in both of these special methods of working is that concerned with the balancing of strain from two or more workings.

Partial extraction by working panels and leaving pillars equivalent to approximately a quarter of the depth of the workings has been successfully used in many cases in recent years. This has enabled large areas of coal to be partially worked (usually 50%) which hitherto would have been left and considered to be commercially unworkable.

Conclusions

The rights to work coal and to withdraw support from the surface together with the liability for compensation are subject to statute and common-law provisions. There is also overall control by planning legislation. However, there now exists a sufficient understanding of the practical effects of mining subsidence to enable the respective developments of surface and mineral estates to take place without great conflict, provided that proper consultation and co-operation occurs between the parties concerned. Successful arrangements exist in many localities and future liaison will be of increasing importance, especially in areas of concentrated development.

Acknowledgements

The author is grateful for the National Coal Board's permission to publish this document; the views expressed, however, are those of the author and not necessarily those of the Board. Figs. 6.6, 6.8 and 6.9 are reproduced from the Subsidence Engineer's Handbook[3] by kind permission of the National Coal Board.

REFERENCES

1. Halbaum, H. W. G., 'The Action, Influence and Control of the Roof in Longwall Workings,' *Trans. Inst. Min. Eng.* (1903)
2. Briggs, H., *Mining Subsidence,* Arnold, London (1929)
3. *Subsidence Engineer's Handbook,* National Coal Board, London (1966)
4. Wardell, K., 'Some Observations of the Relationship Between Time and Mining Subsidence', *Trans. Inst. Min. Engrs.,* 113, 471 (1953/1954)
5. Orchard, R. J., 'Recent Developments in Predicting the Amplitude of Mining Subsidence', *J. R. Inst. Chart. Surv.,* 33, 864 (1954)
6. Mather, J. D., Gray, D. A. and Jenkins, D. G., 'The Use of Tracers to Investigate the Relationship Between Mining Subsidence and Ground Water Occurrence at Aberfan, South Wales', *J. Hydrology,* 9, 136–154 (1969)
7. Moseley, F. and Ahmed, S. M., 'Carboniferous Joints in the North of England and Their Relation to Earlier and Later Structures', *Proc. Yorks. Geol. Soc.,* 36 No. 1, 61–90 (1967)

Chapter 7

Characteristics of Shallow Coal-mine Workings and Their Implications in Urban Redevelopment Areas

In this chapter the behavioural characteristics of shallow mine workings and their evaluation at the planning stage of site investigations for urban redevelopment schemes is considered. Emphasis is placed upon the behaviour of room and pillar workings which present the most intransigent problems because it is particularly difficult to quantify their long-term behaviour.

It should be appreciated that mine workings relating to underground exploitation of other rocks and minerals besides coal, fireclay-ganister and ironstone, can be expected in coal-bearing strata in the UK. For example, oil shales in the Lothians and parts of Fifeshire, limestone for steelmaking, and even more mundane extractions such as flagstones (Fig. 7.1) should be anticipated, even though a particular part of the coal-bearing sequence may to all intents and purposes appear to be barren.

Shallow workings close to outcrop should be expected in any urban area where exploitable beds (particularly well-documented horizons) are not covered by thick superficial deposits. Moreover, it must be stressed that the Geological Survey maps and Memoirs, which are a useful guide at the pre-site investigation stage, may not be strictly accurate in the older towns and cities. This is because the urban expansion of the industrial revolution had already obliterated much useful information by the time mapping commenced in the last century. Furthermore, the first obligation to keep mine records dates from 1850 and it was not until after about 1872 that mine plans were compulsorily deposited. Hence, from a mining view point also, information may well be sparse for those industrial towns which expanded rapidly towards the end of the last century.

With large urban redevelopment schemes it is imperative that the site investigations be phased, and design concepts and structural configurations remain flexible, so that due consideration can be given to the impact of shallow mine workings on the proposed scheme.

Types of Workings

BELL-PITS

Mediaeval bell-pits put down to exploit clay–ironstones are a feature of coalfield areas where the drift cover is thin. This type of working (Figs. 7.2 and 7.3) was subsequently adopted for the mining of coal as well and was commonly used until the seventeenth century. In general the shafts rarely exceeded 12 m in depth and their diameter was usually about 1·2 m. Radial mining from the bottom of the shaft continued until the area of extraction was such that natural or artificial support was not feasible. A whole progression of bell-pits may often delineate the outcrop of a seam and in open country cones of waste around the shafts may still be visible.

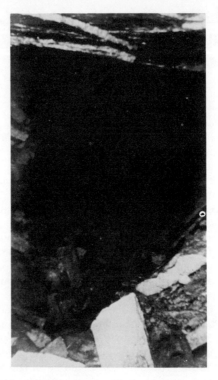

7.1. Entrance to gallery driven into Elland Flags at Gamble Hill, near Bramley, Yorks. [Grid Ref. SE 253 334]. The 2·1 m high by 1·5 m wide workings were at two horizons and may have been up to 152 m long (after Taylor[1])

There is evidence that the state of compaction of very ancient bell-pits may be far from satisfactory[1], and like conventional mine shafts they may prove difficult to locate.

Good sample recovery is necessary to elucidate the presence or otherwise of back-fill and the extent of the working. It is pertinent to mention that on a recognised sampling frequency (CP 2001 : 1957)[2] soft ground shell and auger holes failed to detect partially collapsed bell-pits under only 1·5 m of cover at Heath Estate, Leeds [Grid Ref. SE 283 308]. A subsequent rapid 'open-hole' survey using a rotary air-flush rig with tri-cone roller bit enabled the geological structure of the coal outlier and nature of the overburden to be assessed. Utilising foundation exposures of the 2·4 m thick Beeston Bed Coal, together with a fully cored hole to the deep of the area, an open-cast proposition of 21 740 Mg of coal was shown to be viable at a ratio of 8·2 (overburden) : 1 (coal). Time prevented the full

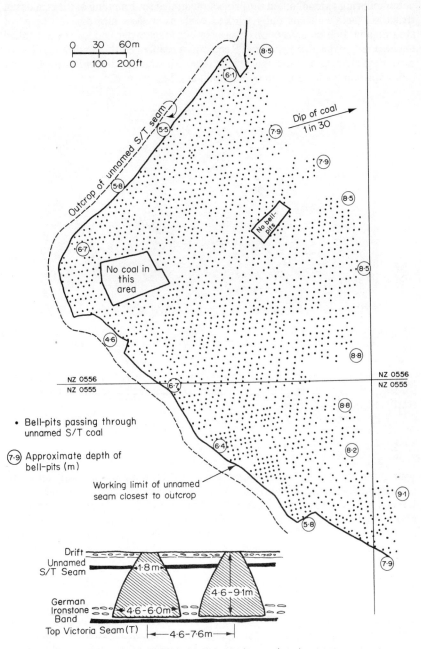

O   30   60m
O   100   200ft

Outcrop of unnamed S/T seam

Dip of coal
1 in 30

(8·5)
(6·1)
(5·5)
(7·9)
(5·8)
(7·9)
(8·5)
No bell-pits
(6·7)
(8·5)
No coal in this area
(4·6)
(8·8)

NZ 0556
NZ 0555

NZ 0556
NZ 0555

(6·7)
(8·8)
(6·4)
(8·2)
(9·1)
Working limit of unnamed seam closest to outcrop
(5·8)
(7·9)

• Bell-pits passing through unnamed S/T coal

(7·9) Approximate depth of bell-pits (m)

Drift
Unnamed S/T Seam
German Ironstone Band
Top Victoria Seam (T)

1·8 m
4·6-9·1m
4·6-6·0m
4·6-7·6m

Relationship of bell-pits to exploited ironstone horizon and coal seams

7.2. Unique intensity of bell-pits in German Ironstone exposed at Sproats Opencast Site, Northumberland. The pits which were well backfilled have the characteristic 'bell' shape without any shaft section (courtesy *The National Coal Board Opencast Executive*)

potential being exploited, but the project facilitated the founding of the remaining structures below the partially worked coal, as well as affording a degree of landscaping. Bell-pit extractions averaged 7 m in diameter and some were linked together by spider-like headings. Remedial measures of this nature involve an extra time element, but they have been used expeditiously in a number of recent motorway contracts as noted by Shadbolt and Mabe[3].

7.3. Excavated bell-pit working in German Ironstone Band at Sproats Opencast Site, Northumberland. Bell-pits generally 4·6–6·1 m in diameter at base

Occasionally conventional geophysical methods prove successful in locating the shafts, but like aerial photograph interpretations their usefulness is obviously restricted in built-up areas.

ROOM AND PILLAR WORKINGS

With this method (Fig. 7.4a) pillars of coal are left to support the overlying strata and essentially the method is the second stage in the evolution of coal mining in the UK. Cameron[4] cites very small stoops (about 1·2–1·8 m square) at intervals of about 6–9 m as being symptomatic of the earlier days of mining. Atkinson[5] also refers to pillars about 3·3 m^2 but with rooms some 2·7 m wide, as the working dimensions of the early seventeenth century. By the middle of the eighteenth century pillar sizes had increased to about 5·5 m by 3 m with rooms about 3 m in width. Generally, room widths of between 1·8 and 4·6 m would appear to be the norm prior to the last century[6]; coal extraction accounted for between 30 and 70% of the areas worked. In countries like the USA where the method has been extensively used in recent times, little information has been collated and published. It is of interest to note that the design of this type of system is still a matter for debate[7,8].

Special types of working methods such as 'Staffordshire square work', appertaining to the Thick Coal of abnormal thickness (up to about 9 m) were developed in some coalfields to suit local conditions. Some of the earlier total extraction techniques, like the panel workings of Thomas Barnes in Durham (1795), were to some extent a combination room and pillar and longwall approach.

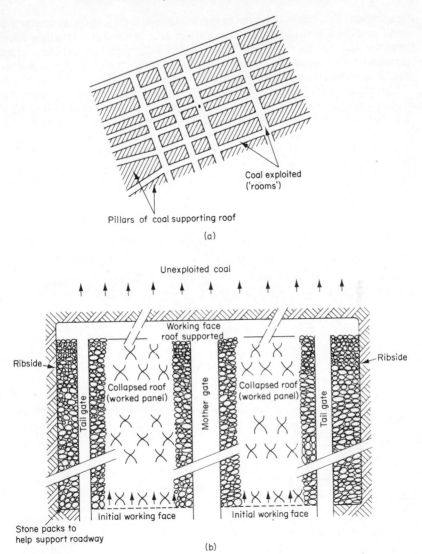

7.4. Diagrammatic representation of two common methods of coal extraction, (*a*) plan view of room and pillar and (*b*) plan view of double longwall advancing face (after Attewell and Taylor[33])

LONGWALL WORKINGS

Longwall workings (Fig. 7.4*b*) were developed in the late seventeenth century and are generally the modern method of mining in the UK. The panels are developed from an initial drivage within the seam and as the face advances supports are withdrawn, which allows the roof to collapse behind the current working face.

Exploitation by this method produces surface subsidence which decelerates rapidly and tends to reach equilibrium about a year after extraction within the critical area has occurred[9]. Extraction of shortwall pillars during the retreat phase of the room and pillar mining simulates the longwall surface condition, although it can never be assumed that all pillars have been removed.

The detailed field data pertaining to surface behaviour, brought together in the *Subsidence Engineers Handbook*[9], enable predictions to be made which have credence with respect to subsequent measurements. Predictions are usually correct to within ±10% in most cases. For the proposed Tyne sewerage syphon tunnel, cumulative tensile strain contours which are a reflection of extractions in the underlying High Main and Maudlin seams, were constructed by J. B. Boden at Durham University (Fig. 7.5). The assessment of the mass behaviour of the Coal

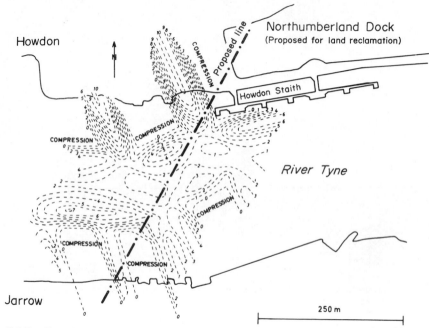

7.5. Tensile strain contours (mm/m) for workings beneath the River Tyne [Grid Ref. NZ 334 660]. High Main 180 m below rock-head (north bank), 155 m (south bank); Maudlin 256 m (south bank)

Measures rocks along the line of the tunnel as a consequence of induced tensions and compressions is one of the factors from which first approximations can be made at the planning stage with respect to stability and water seepage during subsequent driving operations.

It should be mentioned that implicit in the empirical subsidence predictions[9] is the fact that the geology of the superincumbent strata is far less important than the geometry and dimensions of the workings. In the writer's opinion geology (in terms of lithology and structure) has a much more important influence on the ultimate behaviour of room and pillar workings, as will become evident later in this chapter. Suffice it to say that the *Subsidence Engineers Handbook*[9] is not applicable to room and pillar workings, apart from total extraction assumptions.

Mine Shafts

The precise location of shafts (or an attempt to eliminate the possibility of their presence) is of prime importance to the safety of a potential structure. Similarly, from an economic standpoint land sterilisation is unrealistic and large areas should not be excluded on a probability basis without any attempt to locate suspected shafts. The problem can be categorised under two headings, (*a*) their location and (*b*) the method, nature and condition of in-fill (if present at all).

From earlier comments it can reasonably be accepted that prior to 1872 there are few records of shafts put down during the previous 200 years. Dean[10] discusses the hazards of old shafts, citing 19 case histories. He also highlights the size of the location and collation exercise in referring to the National Coal Board's west Lancashire Area, where 3000 shafts are known to exist, though not necessarily in their precise locations. On the 1884 Ordnance Sheet for the Netherton area of Dudley, for example, some 65 shafts were already documented (commonly in pairs) within an area of just over 1 km².

Fortunately many old shafts are of relatively small diameter or size (Fig. 7.6), so the frequency and extent of failures tend to be limited. However, old shafts should be treated as a hazard until properly investigated. It is important to recognise that

7.6. One of two shafts encountered during construction of A1 viaduct at Gateshead [Grid Ref. NZ 25766 63200]. The unlined (well-finished) shaft was 1·4 m in diameter and in-filled, and subsequently covered with a reinforced concrete raft

the condition of abandonment may entail open shafts, capping at the surface, or, as was often the case, haphazard in-fill to a level at which a wooden raft was possibly 'needled' across the void. Stopping-off of shaft insets prior to complete infilling is always difficult to ascertain with older workings and settlement of the in-fill is a common feature even though the lining may be in excellent condition. Collard[11] describes this situation in relation to the new science library of Durham University. It should be noted that of the six shafts referred to by Collard, only two are shown on the ordnance sheets and some of the existing mine plans (Fig. 7.7). If, as in the above case, the shafts can be accurately investigated and fixed, proposed structures can sometimes be advantageously positioned so as to be protected by the coal pillars that protected the shafts, provided the pillars have not been subsequently mined out.

When locations can be fixed approximately from ordnance sheets or old mine plans, the question of accurate positioning becomes paramount. Geophysical methods such as proton magnetometry have on occasions proved useful[12], but

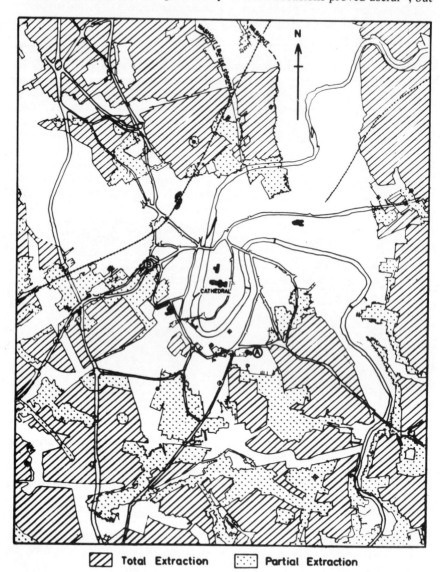

� Total Extraction � Partial Extraction

⊕ Old Shafts

7.7. Coal extraction beneath Durham City from the Hutton seam. *A* two shafts only on University science site; *B* area subject to subsidence in 1906. The other principal workings in underlying Busty seam lie almost entirely to the west of a N–S line through the peninsula. Centre of city virtually free from workings

confirmation by excavation is always necessary. This objective can prove frustrating if the area is covered with a mantle of colliery discard—such was the situation at Eslington Road, Gateshead, when an attempt was made to locate shafts by excavation alone. Eventually two shafts were located by drilling, one being plugged at about 2·5 m below ground level by sound reinforced concrete, whilst the other was plugged with timber at about 9 m below ground level and a water loss of 91 000 litre was recorded during drilling.

A useful practice in assessing likely shaft locations is to fix their position from as many plans as possible—sequential Ordnance Sheet editions and abandoned mine plans for the respective seams worked. By drawing equivalent circles of likely dimensions, the overlap area provides a probability zone for detailed examination. It is always useful to allow for both rotary- and soft-ground boring rigs to be available for this type of investigation; the former for forward penetration and the latter for sampling and Standard Penetration tests.

Collapse Mechanism of Room and Pillar Workings

In coal-bearing strata the roof rocks may vary from weak, thinly-bedded, marine shales to strong limestones or sandstones. Similarly, seatearths beneath coal seams may vary from plastic fireclays to strong carbonaceous mudstones and siliceous ganisters. The intrinsic strength of the coal* varies, but what is more important in the context of pillars is that their ultimate behaviour is a function of seam thickness to pillar width, the depth below ground level and the size of the extraction area[7]. Wardell and Wood[6] noted that as a rule when the width of the span is less than one fifth the depth of overburden then it will remain self supporting.

The question to be resolved is the mode of failure of this type of working which must consequently concern the roof rock lithology, the coal itself and the nature of the floor materials. The wholesale collapse of pillars on a large scale in the UK is apparently not a significant feature (c.f. Coalbrook, S. Africa; Bryan *et al.*[13]). Localised pillar failures under low overburden pressures have recently been reported from the USA[14,15]. Gray and Meyers[14] noted that the pillars in the 1·8 m thick Pittsburgh Coal at a depth of 33·5 m below ground level were probably too small (3 m in width), with the rooms about 6 m wide (75% extraction), whilst Mansur and Skouby[15] suggested that the partial pillar failures and subsidence at Belville, Illinois, could well have been activated by a moderate earthquake in 1965.

It must always be borne in mind that pillar robbing (i.e. subsequent cuts taken off the top of one or more sides of pillars) can reduce their strength, so leading to stress redistribution and failure. *In-situ* investigations following subsidence in Durham City in January 1906 attributed the cause to workings in the 1·1 m thick Hutton seam, at some 55–67 m below ground level (*see* Fig. 7.7). Pillar robbing had increased the extraction percentage from the permitted 30 to nearer 80 in the area concerned.

Slow deterioration and failure of pillars may take place after years have elapsed[9], although observations at shallow depth on open-cast sites and the resistance of coal to low temperature oxidation recorded in colliery tip investigations, infer that failures due to coal weathering alone are uncommon. Cameron[4] concluded that any remaining small pillars from the early days of room

---

* Modern work proposes that pillar strength can be assessed in terms of the strength of an outer yielding 'skin', together with the strength of the pillar 'core'. Weak dirt bands within coal seams are another important consideration.

and pillar workings will have crushed out at depths of 46 to 61 m. Based on empirical rules such as those of Orchard[16], this contention is not unreasonable. Orchard suggested that stability can be ensured if the pillar width to depth ratio is more than 0·1. Alternatively, it can be demonstrated from beam analogies (see below) that in the absence of discontinuities self-supporting spans are dependent on the width between pillars (rooms) being less than about 0·2 to 0·25 of the depth to cover.

Failed pillars are occasionally observed at very shallow depth in the vicinity of faults which act as safety valves[17], the consequent readjustment 'steps' giving rise to differential subsidence which may well damage existing structures[18]. Subsequent workings at greater depth undoubtedly accentuate this phenomenon. The Institution of Civil Engineers[19] recommends that structures should preferably not be erected within 15 m of either side of known fault surface locations.

Generally, the dimensions of pillars are usually such that at shallow depth their stability is unlikely to be further reduced by structural surcharge.

Estimates of the average stress on pillars due to superincumbent strata can be determined approximately as shown in Table 7.1. Some authorities (e.g. Wardell and Wood[6]) add the full structural surcharge to the overburden pressure in order to determine an overall stress on the pillars. It is likely that the additional load due to buildings will be spread laterally to some extent however, and the example in Table 7.1 shows the variation with depth when the stress distribution is assumed to be conventionally spread in the ratio 2 (vertically) to 1 (horizontally). An alternative approach is to consider the arching theories mentioned below, basing the increase in stress on the theoretical term pertaining to surcharge (Table 7.1 and Terzaghi[21]).

When mine plans and other requisite data are available it is sometimes feasible to attempt to calculate a factor of safety against pillar failure by dividing pillar strength by the stress on the pillars (overburden stress plus additional structural surcharge). The sensitivity of the various parameters can be assessed by considering the best and worst conditions. A number of techniques can be adopted for assessing the strengths of pillars from laboratory and *in situ* tests on coal[24,25]; the general orders of magnitude for the strength of a few small sized pillars referred to in the literature are given in Table 7.1.

Instability may also be induced by heaving of the seatearth floor with the material completely infilling the rooms. This rheological behaviour of wet seatearth is marked in certain seams, one probable factor being the presence of expandable montmorillonite as a mixed-layer clay component[26,27].

Irrespective of the aforementioned processes, each one of which may well have a bearing on the stability of the workings (and should be carefully considered at the planning stage and during the site investigation), the most common form of instability involves failure of the roof strata into the available rooms. Methods of analysis of this failure mode are numerous because of subsidence implications associated with mining and tunnelling. The overlying strata have been treated as a continuous elastic body, as a beam or plate, and pressure arch and dome theories have been developed and reviewed by numerous authors (Terzaghi[21], Denkhaus[28], Szechy[29] to name but a few).

The prominence of bed separation in fissile Carboniferous rocks has led to the evaluation of theoretical maximum distances for the rocks which span a room and the development of mechanical arches (e.g. Briggs[30]). By equating the maximum bending moments, the span length L can be expressed as:

Table 7.1 COAL PILLARS—RELATIVE STRESSES AND STRENGTHS

Assume: foundation area = 15 m × 15 m, i.e. $2B = 15$ m; contact pressure due to structure $(q) = 295$ kN/m² (relatively high loading); percentage of old workings = 70, $e = 0·7$; density = 2·308 Mg/m³ (based on conventional stress of 1 lbf/in² per foot depth); $\phi = 45·5°$ (Spears and Taylor[20])

Depth (z) to workings (m)	Stress due to superincumbent overburden [(depth × density) ÷ (1 − e)] (kN/m²)	Stress due to structure 2:1 spread load ÷ (1 − e) (kN/m²)	Arching* (kN/m²)	Strengths of small pillars (kN/m², approx.)	Pillar size width (m)	Pillar size height (m)	Author
5	377	553	499	5500	1·5	0·9	Szeki[22] in situ tests:
10	755	354	253	6900	7·6	0·9	computed using
15	1132	246	128	2400	1·5	3·7	conventional empirical
30	2264	109	17	3100	7·6	3·7	formulae
				3490–8270	small pillars		Greenwald, Howarth and Hartmann[23]
				4480	1·5 m cube		Bieniawski[24]

* Additional structural stress (adapted from Terzaghi[21]) = $\dfrac{q\exp(-K\tan\phi\, z/B)}{1-e}$ where $K = 1$ (at least); other terms defined above.

$$L = \sqrt{\frac{2\sigma T}{\gamma}}$$

where σ = tensile strength of the 'rock beam',
     T = thickness of the beam and
     γ = bulk density of the rock.

Obviously, it is difficult to quantify this type of approach because of lithological, discontinuity and strength variations expected in cyclothemic sequences. In other words, the precise nature is likely to be over-simplified, particularly as the original void height, dip of beds and bulking characteristics of collapsed material are also important behavioural controls. Observations of arching and void migration are demonstrated in Figs. 7.8 to 7.10.* In practice, progressive collapse of lithologically similar rocks spanning a void will lead to an upward migration of the void (Fig. 7.9). The collapse ceases when an overlying bed possesses a sufficiently high tensile strength and competency to preclude further migration (Fig. 7.11), or when closure results in a degree of self-support (Fig. 7.8). Exceptionally, the void may reach ground level, giving rise to a 'crown-hole'. Two recent crown-holes in Ashtree Gardens, Gateshead (Fig. 7.12) were approximately 1·8 m in diameter. It is believed that these workings (possibly of 1750 vintage) in the 1·7 m thick High Main seam are some 9·0 to 12·0 m below ground level at this location. Ingress of saturated unconsolidated deposits into underlying voids can greatly increase the area of collapse.

A sudden failure resulting in a crown-hole about 5 m in diameter occurred during the prospecting phase of Mill Moor open-cast site (Fig. 7.13). This was a joint-controlled failure of the sandstone overlying the Black Band seam and was possibly accelerated by air-flush boreholes ($480 \, kN/m^2$ pressure) disturbing the incipient equilibrium. The collapse above the underlying Whinmoor workings had probably been attenuated some 8·8 m above the seam (about 8 or 9 times the coal thickness).

The height to which a void might migrate, or collapse take place, is of the utmost concern at the planning stage of a site investigation. With arenaceous competent rocks the beam may effectively span the void, but natural failure of bedded arenaceous and silty rocks can take the form of a joint-controlled quasi-cantilever-type failure. Importantly, isolated occurrences such as Mill Moor demonstrate clearly that seemingly competent, lithified rocks with widely spaced joints may well be the very types that are subject to sudden joint-controlled collapse, particularly if the original voids are large and migration of material has taken place 'down dip'. Water may play an important role in such circumstances.

Turning to the question of the height above seam level to which a strata arch can close, the writer has noted that in shales and shaly mudstones this is commonly one or two times the width of the intervening room. In many respects this height is compatible with the concept of the pressure arch or dome and it is of interest to record that Ackenheil and Dougherty[31] use a figure of twice the distance between the supports for their approximation of arch development height above a seam.

Although possibly more applicable to longwall workings, the more exceptional cases in which discontinuities promote almost vertical failures adjacent to the pillars can be investigated semi-quantitatively. One is faced with the difficulty of

* The arch in the brick wall 'model' is analogous to an infilled and cemented jointed rock condition.

7.8. Excavation showing arch closure beneath about 12 m of shale cover at Pethburn Opencast Site [Grid Ref. NZ 180 480], Co. Durham. Working in 2·4 m thick Five-Quarter seam; distance between pillars 4·9 m; height of arch 7·62 m

determining the increase in volume (decrease in density) that could occur in order that the void and developing arch may be infilled. In many ways colliery discard which contains all the rock types likely to be found in collapsed workings is not an unreasonable analogy from which to obtain an 'average' value for volumetric increase. Using the modal density value for colliery tips (2 Mg/m³), Tincelin's[32] expression is fitting (*see also* Wardell and Eynon[7]):

$$H = t \left[\frac{\gamma_1}{\gamma} \middle/ 1 - \frac{\gamma_1}{\gamma} \right]$$

where H = total height of collapse,
t = thickness of the seam,
γ_1 = bulk density of collapsed roof materials (e.g. 2·00 Mg/m³),
γ = bulk density of the roof rocks (2·24 Mg/m³*).

7.9. Void migration above Five-Quarter seam at Pethburn Opencast site. Five-Quarter seam has collapsed into workings in the underlying Main seam; migration height about 3·7 m

7.10. Arch development in an old, 3·7 m high, garden wall in Co. Durham

H may therefore conceivably be equal to about $8t$, which is generally of the right order for many crown-hole observations, and may be taken into account as a first approximation in the planning stages of a site investigation.

The effects of mining on the opening up of discontinuities in overlying younger rocks such as the Magnesian Limestone has been commented upon by Shadbolt and Mabe[3]. Pertinent to their comments and the concept of beam behaviour is the incidence of 'pit falls' at Houghton-le-Spring, investigated by M. J. Turner (*see* Attewell and Taylor[33]). These large tension gashes in the Lower Magnesian Limestone have an alignment roughly parallel to outcrop (Fig. 7.14). The oldest gashes are closest to outcrop, the newer ones lying progressively further to the deep beneath the crest of the slope. Basing the analysis on the analogy of a cantilever beam pinned at one end, cross-sections of the rock outcrop were divided into their constituent elements (Fig. 7.14). Laboratory tests gave an average tensile strength

7.11. Working in the 0·6 m thick Top Brockwell seam at Sproats Opencast Site, Northumberland. The collapse under about 6 m cover has been attenuated by a bed of flaggy sandstone close to top of seam (courtesy *National Coal Board Opencast Executive*)

7.12. One of the two crown-holes which recently occurred in Ashtree Gardens, Gateshead [Grid Refs. NZ 25765 61224 and 25778 61188]

* Taking a mean value for the Mansfield Marine Band cyclothem.

140

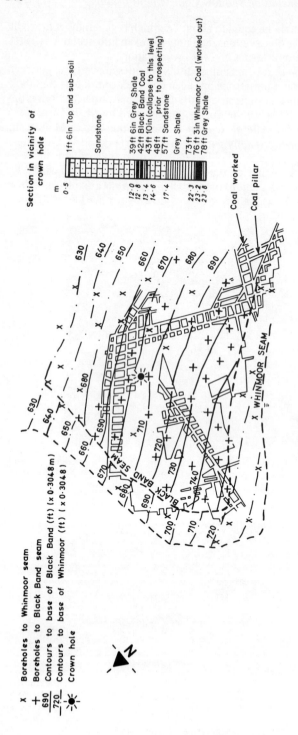

Section in vicinity of crown hole

1ft 6in Top and sub-soil

Sandstone

39ft 6in Grey Shale
42ft Black Band Coal
43ft 10in (collapse to this level prior to prospecting)
48ft Sandstone
57ft Sandstone

Grey Shale

73ft
76ft 3in Whinmoor Coal (worked out)
78ft Grey Shale

Coal worked
Coal pillar

x Boreholes to Whinmoor seam
+ Boreholes to Black Band seam
690 Contours to base of Black Band (ft) (x 0·3048 m)
720 Contours to base of Whinmoor (ft) (x 0·3048)
Crown hole

7.13. Crown hole at Millmoor Opencast site, Western Section, Yorks. [Grid Ref. SE 304 003]. Collapse within an area where Whinmoor had been totally extracted; majority of boreholes penetrated only the upper, Black Band seam (courtesy *National Coal Board Opencast Executive*)

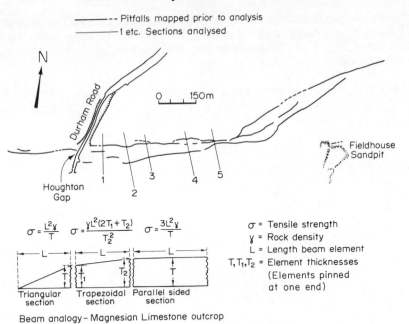

Beam analogy – Magnesian Limestone outcrop

7.14. 'Pit-falls' in the Magnesian Limestone at Houghton-le-Spring, Co. Durham [Grid Ref. NZ 344 505]. Master joints are included within beam elements; back-analysis inferred that other, older 'falls' were present to the south of those surveyed; naturally infilled ground breaks were subsequently recognised adjacent to outcrop

of $1·475$ MN/m^2 and calculation of the equivalent fracture spacing in the field produced surprisingly accurate locations. It is inferred that the intrinsic discontinuities did not sensibly affect the rock's beam-like behaviour when underground support was removed from it and its carpet of Basal Permian Sands near outcrop.

Shallow Workings Beneath Urban Redevelopment Areas

Urban redevelopment is not solely concerned with founding structures near ground level, or transmitting these loads to other deeper horizons of adequate bearing capacity. Deep cuttings for urban roads may reduce an otherwise adequate cover of competent strata above old workings to a precarious level. Similarly, for access and service tunnels (Figs. 7.5 and 7.15) full awareness of possible workings is of the utmost importance.

Mine plans of workings beneath the older urban areas are likely to be sparse; the detailed coverage for the two principal seams worked beneath Durham City, for example (Fig. 7.7) is the exception rather than the rule. It should also be appreciated that the older dwellings often had cellars incorporated, and consequently the loads imposed at foundation level were minimal—there may well be little visual evidence of structural damage due to subsidence, or settlement of in-filled crop-workings. This does not necessarily mean that workings are not present. Probe borings put down recently in the centre of Gateshead when the 1·22 m diameter G.P.O. tunnel encountered partly in-filled, very shallow workings in the High Main seam (Fig. 7.15), demonstrate just how tenuous the current state of equilibrium can be.

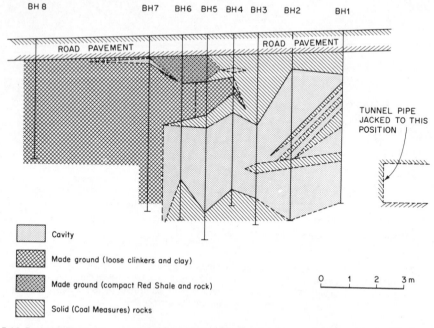

7.15. Probe borehole cross-section showing voids immediately below road pavement in the G.P.O. tunnel, Oakwellgate, Gateshead [Grid Ref. NZ 25610 63446]. Tunnel completed in 'cut and cover'

Details of possible sources of records relating to mining activities and necessary geological information are given by Dumbleton and West[34]; needless to say, this pre-site investigation search is usually time consuming. For the Yorkshire Development Group's two-stage redevelopment at Hunslet, Leeds [Figs. 7.16 (stage 1) and 7.17 (stage 2)] the Estate plan of 1824 showed no direct evidence of any underground mining activities although one of the access routes was designated a 'coal road'. Geological Sheets 218 NW and SW (Yorkshire) implied that the sites lay between the Beeston Bed and overlying Blocking Bed, the dip being in a south easterly direction. The major Osmondthorpe Fault was shown to cross the area (Fig. 7.17), but this was subsequently shown to have died out, or possibly to have split into minor faults. The only known Beeston Bed workings were those of Middleton No. 2. Colliery, dated 1890 (Fig. 7.17) in which the superior 1·2 m of the 2·4 m seam had been worked. However, a photograph by Holgate[35] of workings in the subsequently backfilled Hillidge Road/Jack Lane brick pit (Fig. 7.17) proved conclusively that workings could be expected within the site areas.

The aim of the investigation was therefore to establish the extent of the workings and to determine their state. Indisputable workings were found in only two holes in the northern site area (Fig. 7.16) in the vicinity of the old brick pit. The cored

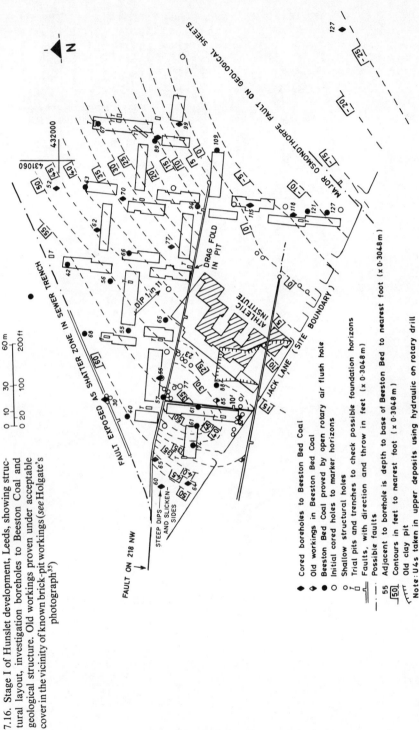

7.16. Stage I of Hunslet development, Leeds, showing structural layout, investigation boreholes to Beeston Coal and geological structure. Old workings proven under acceptable cover in the vicinity of known brick-pit workings (see Holgate's photograph[35])

◆ Cored boreholes to Beeston Bed Coal

◆ Old workings in Beeston Bed Coal

● Beeston Bed Coal proved by open rotary air flush hole

○ Initial cored holes to marker horizons

○ Shallow structural holes

☐ Trial pits and trenches to check possible foundation horizons

---- Faults, with direction and throw in feet ($\times 0.3048$ m)

—·— Possible faults

55 Adjacent to borehole is depth to base of Beeston Bed to nearest foot ($\times 0.3048$ m)

50 Contours in feet to nearest foot ($\times 0.3048$ m)

Old clay pit

Note: U4s taken in upper deposits using hydraulic on rotary drill

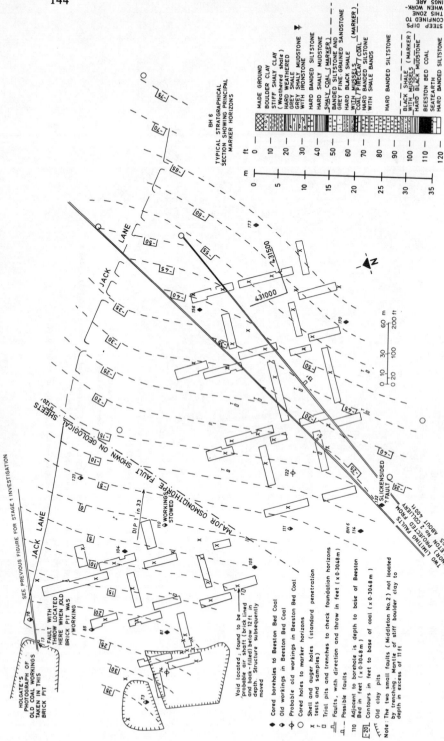

7.17. Stage II of Hunslet development. Showing *initial* structural layout which took into account the major

TYPICAL STRATIGRAPHICAL SECTION SHOWING PRINCIPAL MARKER HORIZONS

BH 6

MADE GROUND	
BOULDER CLAY	
STIFF SHALY CLAY	
HARD WEATHERED)	(Weathered shale)
GREY SHALE	
GREY SHALY MUDSTONE	
HARD	WITH IRONSTONE
HARD BANDED SILTSTONE	
HARD SHALY MUDSTONE	
SHALY COAL (MARKER)	
BANDED SILTSTONE AND	
GREY FINE-GRAINED SANDSTONE	
HARD BLACK SHALE	
WITH MUSSELS	(MARKER)
COAL FIRECLAY COAL 7 COAL	
HARD BANDED SILTSTONE	
WITH SHALE BANDS	
HARD BANDED SILTSTONE	
BLACK SHALE (MARKER)	
WITH MUSSELS (MARKER)	
HARD BLACK MUDSTONE	
BEESTON BED COAL	
SEATEARTH	
HARD BANDED SILTSTONE	

STEEP DIPS
CONFINED TO
THIS ZONE
WHEN WORK-
INGS ARE
PRESENT

ft   m

Legend:

◆ Cored boreholes to Beeston Bed Coal

◇ Old workings in Beeston Bed Coal

◇ Probable old workings in Beeston Bed Coal

○ Cored holes to marker horizons

× Shell and auger holes (standard penetration
   tests and samples)

□ Trial pits and trenches to check foundation horizons

⌐⌐ Faults, with direction and throw in feet (x 0·3048 m)

⌐⌐ Possible faults

110   Adjacent to borehole is depth to base of Beeston
      Bed in feet (x 0·3048 m)

[20] Contours in feet to base of coal (x 0·3048 m)

Note: The two small faults (Middleton No.2) not located
      by trenching. Marlin of stiff boulder clay to
      depth in excess of 11ft

Void located ; found to be
'probable air shaft (brick lined
and back-filled) below 12ft in
depth. Structure subsequently
moved

SEE PREVIOUS FIGURE FOR STAGE 1 INVESTIGATION

HOLGATE'S
PHOTOGRAPH OF
OLD COAL WORKINGS
TAKEN IN THIS
BRICK PIT

JACK LANE

FAULT WITH
THROW LOCATED
HERE WHEN OLD
BRICK PIT WAS
WORKING

MAJOR OSMONDTHORPE FAULT SHOWN ON GEOLOGICAL SHEETS

WORKINGS STOWED

DIP 1 in 33

JACK LANE

431000

2·31500

BH 6

SLICKENSIDED FAULT

TWO LIMITING FAULTS PROJECTED FROM
MIDDLETON No. 2 COLLIERY ABOUT 400 ft
TO SW

(MINOR FAULTS)

boreholes showed that the stratified roof measures (Fig. 7.17) were in continuity with the rather inferior coal which varied in thickness from $2 \cdot 7$ to $1 \cdot 2$ m. Air-flush drilling at $480 \, kN/m^2$ pressure usually causes a return in adjacent holes when voids are interconnected, but no such returns were noted during the investigation. Confirmation that it was very unlikely that the Jack Lane/Hillidge Road workings were extensive within the northern area was provided by sewerage trench excavations which showed that many of the original back-to-back houses close to outcrop were founded directly on a reduced coal thickness; the roof measures were in continuity with the coal.

Widespread workings were encountered beneath the southern site (Fig. 7.17) with steep dips in the black shale roof measures (without the slickensiding or polishing customarily associated with faults). The absence of voids and the confinement of collapse to the fissile roof rocks implied that the thick cover of overlying siltstones and strong mudstones (albeit open-jointed) had caught up with the initially collapsed shales and attained a state of equilibrium. On both sites therefore, it was reasonable to found the pre-cast concrete structures of cross-wall construction at about 2 to 3 m below ground level using strip foundations. A combination of trial pit inspection, penetration tests and laboratory tests were used to confirm the adequacy of the upper materials as a bearing horizon.

Rotary air-flush boreholes using a combination of 'open' and cored (optimum H size diameter) are the most effective conventional means of establishing the presence and state of workings (*see* later remarks relating to *in-situ* inspection and ancillary equipment). The depths of boreholes are clearly governed by site geology and the nature of the workings. In the writer's experience it has seldom been necessary to drill to depths in excess of 45 m, based initially on empirical considerations already discussed in the previous section. Detailed investigations generally concern workings which are less than 30 m in depth. With respect to the spacing of boreholes, this will depend on the size and flexibility of the redevelopment scheme. For the northernmost site just referred to (Fig. 7.16) it can be seen that the highest intensity of holes is closest to the known workings, but the intensity within the remaining area (in relation to the proposed configuration of structures) was sufficient to preclude a high percentage of workings not being detected. The steadily evolving picture relating to the state and nature of the workings beneath the southerly (deeper) site (Fig. 7.17) facilitated a more curtailed drilling programme.

It must be appreciated however, that in these examples a reasonable degree of flexibility in the lay-out of structures was possible in any case. When the areal extent of the site is small, the nature of the workings difficult to interpret, or it is obvious that remedial measures will be required, borehole spacing will necessarily have to be reduced. Bearing in mind the reservations already considered with respect to the possible behaviour of workings, together with the detail required from the site investigation, it is not uncommon in competent strata for remedial measures to be limited to depths of less than about 18 to 20 m (*see* Cameron[4] for structures; Malkin and Wood[36] for highways and associated works).

It is good practice to determine depth isopachytes and contour to the *base* of the coal in order to elucidate significant geological structures and also to provide a control on the drilling programme. Marker horizons and distinct fossil bands (Fig. 7.17) provide a meaningful control on borehole correlation. Under favourable circumstances it is often possible to use thin coal markers as a datum for coring the bottom sections of boreholes only, in order to reduce costs. Experience has shown

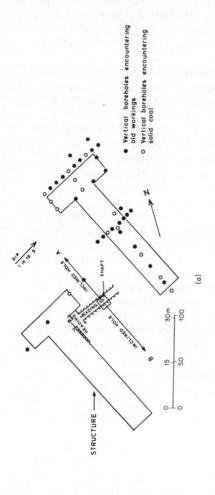

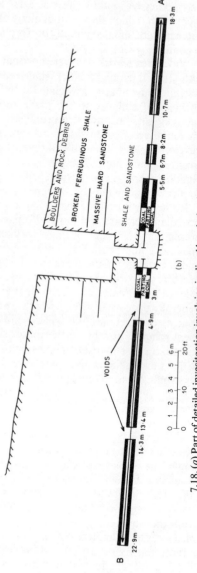

7.18. (a) Part of detailed invesitgation involving inclined boreholes drilled from shafts carried out at Gateshead 18 years ago [Grid Ref. NZ 253 632]; multi-storey structures shown were close to workings in the Metal Coal. (b) Section through inclined borehole A–B.

that boreholes should be put down on an *irregular* grid, and that the numerical incidence of voids to solid coal is not a reliable estimate of workings percentage (this can be adequately demonstrated on a probability basis).

Borehole cameras (which can operate three-dimensionally, *see* Ackenheil and Dougherty[31]) and closed-circuit television may be used for detailed examinations of open or partially collapsed workings. Scale, orientation and illumination place limiting constraints however, on the more conventional equipment.

In-situ examination of shallow workings, shafts and fault zones by trial pits and trenching is invaluable where site conditions permit. Occasionally lateral holes have been drilled from shafts to determine the detailed orientation, density and the current state of the workings (Fig. 7.18). Safety precautions must be rigorously observed when carrying out this type of operation and it must also be borne in mind that disturbance of the ground may create problems at the construction stage.

From the observations made by the writer, and the case histories discussed in this chapter it will be eminently clear that 'on site' supervision and consultation must be given the highest priority.

Acknowledgements

The writer is extremely grateful to the following who have been unfailingly helpful in providing information: Mr. J. L. Hurrell, Borough Planner and Engineer and Mr. A. Marshall, Senior Assistant Engineer, Gateshead. Mr. E. Weston Stanley, City Architect, Leeds. Dr. A. Szeki, Newcastle-upon-Tyne University. The National Coal Board Opencast Executive, particularly the following officers: Messrs. A. B. Mills, J. Roberts and B. H. Bell. Certain of the case histories cited formed part of the project work carried out by the following past M.Sc. Advanced Course students in Engineering Geology at Durham: Messrs. J. B. Boden, A. Marshall, W. H. Robinson and M. J. Turner.

REFERENCES

1. Taylor, R. K., 'Site Investigations in Coalfields: the Problem of Shallow Mine Workings', *Q. J. Engng. Geol.*, 1 No. 2, 115–133 (1968)
2. CP 2001: *Site Investigations,* British Standards Institution, London (1957)
3. Shadbolt, C. H. and Mabe, W. J., 'Subsidence Aspects of Mining Development in Some Northern Coalfields', in *Geological Aspects of Development and Planning in Northern England,* edited by Warren, P. T., Yorkshire Geological Society, Leeds, 108–123 (1970)
4. Cameron, D. W. G., 'Menace of Present Day Subsidence Due to Ancient Mineral Operations', *J. R. Inst. Chart. Surv.* (Scottish Supplement), 19, Part 3, 159–171 (1956)
5. Atkinson, F., *The Great Northern Coalfield 1700–1900,* University Tutorial Press, London, 76 (1968)
6. Wardell, K. and Wood, J. C., 'Ground Instability Problems Arising from the Presence of Old, Shallow Mine Workings', *Proc. Midl. Soil Mech. Found. Engng. Soc.*, 7, 5–30 (1965 for 1966)
7. Wardell, K. and Eynon, P., 'Structural Concept of Strata Control and Mine Design', *Trans. Inst. Min. Metall. (Sect A: Min. Industry)*, 77 No. 743, 125–138 (1968)
8. Voight, B. and Pariseau, W., 'State of Predictive Art in Subsidence Engineering', *J. Soil Mechs. and Found. Div.*, ASCE, 96 No. SM2, 721–750 (1970)
9. *Subsidence Engineers Handbook,* National Coal Board, London (1966)
10. Dean, J. W., 'Old Shafts and Their Hazards', *Trans. Inst. Min. Engrs.*, 126 No. 78, 368–376 (1967)

11. Collard, G., 'Foundation Problems Encountered During the Current University Building Programme', in *Durham County and City with Teesside,* edited by Dewdney, J. C., British Association for the Advancement of Science, Durham, 113–120 (1970)

12. Raybould, D. R. and Price, D. G., 'The Use of the Proton Magnetometer in Engineering Geological Investigations', *Proc. 1st Int. Conf. on Rock Mechanics,* Lisbon, **1,** 11–14 (1966)

13. Bryan, A., Bryan, J. G. and Fouché, J., 'Some Problems of Strata Control and Support of Pillar Workings', *Trans. Inst. Min. Engrs.,* **123** No. 41, 238–254 (1964)

14. Gray, R. E. and Meyers, J. F., 'Mine Subsidence and Support Methods in the Pittsburgh Area', *J. Soil Mechs. and Found. Div.,* ASCE, **96** No. SM4, 1267–1287 (1970)

15. Mansur, C. I. and Skouby, M. C., 'Mine Grouting to Control Building Settlement', *J. Soil Mech. and Found. Div.,* ASCE, **96** No. SM2, 511–522 (1970)

16. Orchard, R. J., 'Partial Extraction and Subsidence', *Mining Engineer,* **123** No. 43, 417–427 (1964)

17. Lee, A. J., 'The Effect of Faulting on Mining Subsidence', *Mining Engineer,* **125** No. 71, 417–427 (1966)

18. Taylor, R. K., 'The Functions of the Engineering Geologist in Urban Development', *Q. J. Engng. Geol.,* **4** No. 3, 221–234 (1971)

19. *Report on Mining Subsidence,* Institution of Civil Engineers, London, 51 (1959)

20. Spears, D. A. and Taylor, R. K., 'The Influence of Weathering on the Composition and Engineering Properties of *in situ* Coal Measures Rocks', *Int. J. Rock Mech. Min. Sci.,* **9,** 729–756 (1972)

21. Terzaghi, K., *Theoretical Soil Mechanics,* Wiley, New York, 510 (1943)

22. Szeki, A., discussion on Wardell, K. and Eynon, P., 'Structural Concept of Strata Control and Mine Design', *Trans. Inst. Min. Met.* (Sect. *A*: Min. Ind.), **77** No. 743, 142 (1968)

23. Greenwald, H. P., Howarth, H. C. and Hartmann, I., 'Experiments on Strength of Small Pillars of Coal in the Pittsburgh Bed', *United States Bureau of Mines Report of Investigations,* 3575 (1941)

24. Bieniawski, Z. T., 'Failure and Breakage of Rock', Ph.D. Thesis, University of Pretoria (1967)

25. Salamon, M. D. G. and Munro, A. H., 'A Study of the Strength of Coal Pillars', *J. S. Afr. Inst. Min. Met.,* **68,** 55–67 (1967)

26. Pearson, G. M. and Wade, E., 'The Physical Behaviour of Seat-Earths', *Proc. Geol. Soc. Lond.,* No. 1637, 24–33 (1967)

27. Taylor, R. K. and Spears, D. A., 'The Breakdown of British Coal Measure Rocks', *Int. J. Rock Mech. Min. Sci.,* **7,** 481–501 (1970)

28. Denkhaus, H. G., 'Critical Review of Strata Movement Theories and Their Application to Practical Problems', in *Symposium on Rock Mechanics and Strata Control in Mines,* The South African Institute of Mining and Metallurgy, 606 (1965)

29. Szechy, K., *The Art of Tunnelling,* Akademiai Kiado, Budapest, 891 (1966)

30. Briggs, H., *Mining Subsidence,* Arnold, London (1929)

31. Ackenheil, A. C. and Dougherty, M. T., 'Recent Developments in Grouting for Deep Mines', *J. Soil Mechs. and Found. Div.,* ASCE, **96** No. SM1, 251–261 (1970)

32. Tincelin, E., *Pressions et Déformations de Terrain dans les Mines de Fer de Lorraine,* Jouve Editeurs, Paris, 284 (1958)

33. Attewell, P. B. and Taylor, R. K., 'Foundation Engineering—Some Geotechnical Considerations', in Dewdney, J. C. (Editor), *Durham County and City with Teesside,* British Association for the Advancement of Science, Durham, 89–107 (1970)

34. Dumbleton, M. J. and West, G., 'Preliminary Sources of Information for Site Investigations in Britain', *Road Research Laboratory Report LR403,* Road Research Laboratory, Crowthorne, Berks (1971)

35. Holgate, B., 'A Description of Six Sections of the Lower Coal Measures', *Trans. Leeds Geol. Ass.,* **14,** 53–58 (1908)

36. Malkin, A. B. and Wood, J. C., 'Subsidence Problems in Route Design and Construction', *Q. J. Engng. Geol.,* **5,** 179–192 (1972)

Chapter 8

Foundations on the Coal Measures

Apart from the conurbations of southern England and parts of Merseyside, the main centres of population in the UK are located over coalfields. This proximity of population and the Coal Measures is a direct legacy of the industrial revolution. In the 18th century and first part of the 19th century, coal was extracted in agricultural areas around the industrial towns. However, in the subsequent period the progressively expanding suburbs spread out over the old mining areas. These suburbs were composed almost entirely of two- or three-storey structures which exert only relatively low loadings on the sub-surface formation. But, in the last two decades, the redevelopment of city areas and the consequent construction of high blocks has re-exposed this old mining activity by highlighting consequential foundation problems. For example, in the Bristol Development Plan Review, 1966, it was recognised that in 'parts of the city land over old coal workings is unsuitable for development with heavy buildings due to uncollapsed workings or shafts'. This situation arose from the encroachment of the suburbs of Bristol eastwards onto the old Kingswood coalfield in the last century. These suburbs, now requiring progressive replacement by modern structures, cannot be redeveloped without careful consideration of foundation conditions. Similar situations can be recognised in many cities elsewhere, such as Sheffield and Newcastle.

In any consideration of foundation design and construction on the Coal Measures, the question of old mine workings generally predominates. However, the properties of the different rock types which make up the Coal Measure cyclothem are themselves variable. For example, the unconfined strength of these rocks, even in an unweathered state, ranges over at least three orders of magnitude from hard siliceous sandstones to weak, uncemented and clayey seatearths. A structure which overlies a dipping sequence may be founded on more than one rock type and uncertainties as to absolute and differential settlement may arise. A further geological aspect of relevance is that, apart from the Bristol coalfield (and even this is controversial), all the exposed coalfields were glaciated. The bedrock was disturbed by ice and can now be hidden by considerable thicknesses of till, sands and gravels, and lake sediments. The consolidation state of these glacial sediments is variable and cannot always be relied upon to provide an adequate foundation for heavy structures. In these circumstances, boreholes, possibly supplemented by shafts or geophysical techniques, provide the only practicable method of foundation evaluation.

Properties of Coal Measure Rocks

One of the most important features of the Coal Measures as engineering materials is their inherent variation in lithology, and hence their response to weathering and deformation. These material property differences can be illustrated, for example, by the uniaxial compressive strengths of small specimens for the four main rock types summarised from the paper by Price, *et al.*[1] in Table 8.1. Several general conclusions can be drawn from this table as follows:

1. There is a considerable total range in uniaxial strength for fresh Coal Measure rocks extending over nearly three orders of magnitude.
2. Within three rock types the range of strength approximates to, or exceeds, one order of magnitude.
3. The specimens tested were those that were amenable to testing and did not break or disintegrate during preparation; strength values are consequently high.
4. These test results were based on intact samples. The corresponding properties for the rock mass would be less in view of the influence of discontinuities.
5. It must be noted that weathering results in a loss of strength and increase in deformability in those materials which are most susceptible to decay (particularly the mudstones/shales and coals).

Table 8.1 UNCONFINED STRENGTH OF COAL MEASURE ROCKS

Rock type	*Range in uniaxial compressive strength* (MN/m^2)
Sandstone	12·3–393
Siltstone	66·6–112
Mudstone/shale	4·8– 73·0
Coal	9·8– 81·0

It will be clear from these observations that the possible range in strength or deformation properties of three or four orders of magnitude referred to earlier is no exaggeration. In particular, weathering may result in a significant deterioration in material properties and, where sample preparation is possible, test results on weathered rocks are very much less than those quoted in Table 8.1. Although laboratory test values are of interest in an appraisal of site conditions and documenting rock properties, the information can only be of limited value taken in isolation from the *in-situ* conditions. A practical approach to this problem involves the use of a point-load strength-testing device (Fig. 1.3) on rock core together with a fracture spacing log. By such means it is possible to log rock in mechanical terms, providing simultaneously an approximate indication of strength and fracturing[2].

The method of assessment of foundation properties available at the present time relies upon an appraisal of site geology based primarily on borehole cores, supplemented by site or laboratory testing of the core. It is conventional also to rely upon the Foundations Code of Practice[3] as a means of recommending appropriate foundation loadings as modified by personal or published experience. Such assessment can be further supplemented by the use of *in-situ* plate-bearing tests or test piles if the foundation uncertainties are great or there is a potential economic justification. Plate-bearing tests are only appropriate where the founding strata can be readily reached by excavation, and without excessive problems of support or ground-water ingress. Under normal conditions such tests would be used on a site where a raft or footings were under consideration as a foundation. A summary table (Table 8.2) has been prepared from two published sources and personal records which summarises the results of 29 plate-bearing tests. In view of the varied form that these tests have taken in terms of plate diameter (150–607 mm) and rate of loading and maximum load intensity (7 to >150 MN/m^2), the results have been simplified by quoting the total secant modulus for the first cycle loading and the stress level at 2·5 mm deformation. In addition, the Foundations Code of Practice[3] has been used, together with the rock descriptions to provide an assessed maximum safe bearing capacity (the weathered rocks have been assessed as soils rather than rocks). It is of interest to note the marked influence of weathering[4] which involves a reduction in modulus of deformation of up to one to one and a half orders of magnitude. There is a general correspondence between the stress levels at 2·5 mm deformation and the assessed maximum bearing capacities.

Foundations on Stable Sites

At sites underlain by Coal Measure rocks which have not been affected by undermining the foundations take the form of either pad or strip footings, or piles. A raft would normally only be used in those circumstances where bedrock was at too great a depth to provide a suitable founding material. The choice between footings and piles is determined by the form of the building sub-structure, ground-water conditions, and the depth and type of overburden. Bulk excavation techniques can provide for economical removal of overburden and weathered rock above the water table to depths of at least 7 m. Below this general excavation level, it is normally more economic to construct piles. Localised excavation of footing bases, again providing ground-water problems are not severe, can be carried out directly from the ground surface or from a broad foundation level created by bulk excavation. Such footings are, in most cases, founded on weathered rock materials. The design of this foundation-type allows for a load-bearing column supported by a pad of larger plan area. Maximum column loads are normally of the order of 5 MN/m^2 which, when distributed into the pad of greater cross-sectional area, are reduced to a stress level on the foundation in the range of about 0·5–1 MN/m^2. Thus, for practical purposes, it is adequate that a foundation can be assessed as having a minimum safe bearing capacity within such a range. Even for large or heavy structures the loading directed through footings onto a rock foundation need not be high. For example, the foundations of the south pier of the Forth road bridge and the main structure and chimney of the Longannet power station (on the River Forth) which are both sandstone, are loaded at 1·4 and 0·7 MN/m^2, respectively. Only the most severely weathered shales and mudstones at rockhead would be incapable of adequately supporting a load of 0·5 MN/m^2. The typical depth of the most extreme weathering of Coal Measure rocks may be about 5 m, although

Table 8.2 SUMMARY OF RESULTS OF PLATE BEARING TESTS: BASED ON MEANED DATA (after Meigh,[5] and Price, Malkin and Knill[1])

Rock types and condition	Number of tests	Secant modulus of deformation (MN/m²)	Stress level at 2·5 mm deformation (MN/m²)	Assessed bearing value (CP 2004) (MN/m²)
Sandstone	8	815	5	4
Weathered sandstone	2	39	0·6	0·3–2
Mixed mudstone/shale/siltstone	7	380	3·9	2
Weathered mixed shale/mudstone/siltstone	2	7·5	0·2	0·3–0·6
Shales and mudstones	6	150	1·3	2
Weathered shales and mudstones	4	15	0·6	0·2–0·4

some weathering effects commonly penetrate to depths of 8–10 m. Under these circumstances, if the rockhead is composed of weathered shales, it is normally possible to penetrate into sounder materials by a few metres of excavation. Many sites are underlain by a dipping sequence in which more than one rock type is present. Under these circumstances it is a relatively straightforward matter to design the depth or size of the footings to cope with different foundation properties. Adjustments to the depth of the footings can also be made at the excavation stage when the foundation is exposed.

The design of piles is determined by a number of factors including the foundation properties, the properties and depth of overburden, ground-water conditions and structural requirements. Column loads in the structure will determine both the size, and arrangement, of the piles. The depth of penetration of the pile into bedrock is a function of the carrying capacity of the pile and the foundation properties. The pile loading is carried by a combination of end-bearing pressure and shaft adhesion. A pile which is resting on the rock surface is an end-bearing pile and, in current practice, is driven or drilled through soft overburden to rockhead. End-bearing piles resting on rockhead composed of Coal Measure rocks of variable lithology could be subject to differential settlement. In this context it is of interest to refer to experience with the foundations for the New Scotswood bridge, Newcastle-upon-Tyne, where mild steel 'H'-piles, carrying up to $1 \cdot 2$ MN working load, were driven through overburden onto sandstones and shales. The mean surface loading could have been of the order of 10 MN/m^2 although the localised contact stresses must have been significantly higher. Smith[6] refers to the appearance of minor cracks in some of the reinforced concrete piers 'presumably due to a slightly uneven pile movement under load'.

Penetration into the rock provides a rock socket, thus increasing both the acceptable bearing pressure and providing shaft adhesion between the concrete of the pile and rock. It is common practice for the maximum safe bearing capacity of a foundation to be increased by 20% if the foundation depth is greater than 600 mm, up to a maximum increase of 100% at $1 \cdot 8$ m depth. When applied to piles this empirical approach relies on both end-bearing action and the resistance to yielding offered by the rock mass around the pile base.

However, in the assessment of the appropriate loading for a pile shaft, adhesion has to be taken into account, although there are relatively limited data available on shaft adhesion between a bored pile and the borehole wall. Nonetheless it is apparent that shaft adhesion can, in a long rock socket, even in shale for example, account for the majority of the load carried by the pile. Figures ranging from $0 \cdot 1$ to $0 \cdot 4$ MN/m^2 have been quoted as appropriate values for shaft adhesion in a spectrum of rock condition from fragmented shale to fissured, hard sandstone[7]. Crutchlow[8] has suggested, alternatively, that the ultimate shaft friction is about a twentieth of the ultimate end-bearing capacity which (in practical terms and using an adequate factor of safety) provides a similar range of figures to those suggested by Thorburn[7]. Significantly higher figures of shaft adhesion can be implied from some *in-situ* tests on piles and caissons and it must be accepted that quoted figures for shaft adhesion are conservative. There is justification in such an approach being adopted in that uncertainty may exist as to the real effectiveness of the rock–concrete bond, particularly in bored piles taking into account the constructional difficulties which may exist. In cases of such uncertainty there are advantages in carrying out a double test in a pile hole[7] which would permit both the end-bearing and shaft-adhesion components to be measured separately.

From a constructional viewpoint, large-diameter piles rock-socketed into their foundations can carry relatively high allowable loadings. For example, a 0·91 m diameter pile penetrating a depth equivalent to two diameters into a fractured shale should, at least in practical terms (maximum safe bearing capacity of 0·6 MN/m² and shaft adhesion of 0·1 MN/m²), carry a load equivalent to the maximum allowable stress in a concrete pile of 5·1 MN/m². The uncertainties which arise in the design of a pile system are often more related to the risk of differential settlement. Such settlement could arise if the pile base was founded on a hard, but thin obstruction, such as a sandstone layer above a soft shale or where groups of piles were sited on rocks of differing properties. Both these situations can occur in Coal Measure rocks and it is for such reasons that an adequate factor of safety has to be provided in the design of piled foundations.

Foundations Underlain by Shallow Mine Workings

If a site for a proposed development is underlain by old mine workings there are a number of alternative approaches which may be adopted:

1. Location or design of the structures so that the adverse ground conditions are avoided.
2. Construction over old workings shown to be stable.
3. Construction of foundations below or outside the limits of the unstable sub-surface workings.
4. Ground treatment including backfilling of open excavations, possibly combined with grouting of ground which has been subject to collapse.

Decisions as to which approach is the most appropriate to adopt can only be based upon investigation, careful evaluation of the site conditions[1,9] and appraisal of the potential instability of the site, and the economic implications of the alternatives. One of the most intractable problems is the uncertainty which exists with regard to progressive deterioration of workings and consequential risk of settlement, particularly in relation to a new development which might radically alter a quasi-stable situation. For this reason, solutions tend to err on the conservative side and this interpretation is supported by the general lack of data on damage arising from old mine workings. Nevertheless at least one case has been recorded of ground collapses when modern terrace housing had been built over old workings in Chalk (Fig. 1.11).

AVOIDANCE OF OLD MINE WORKINGS

The simplest method of coping with shallow workings is to locate the structures in an area that has not been undermined. It is, however, frequently difficult to establish with certainty that working has not occurred at some time in the past. Price[10] has quoted an interesting case from Wigan where seventeenth century workings were on record as having been abandoned once the water table was encountered. On this basis it was possible to demonstrate that the coal seam had never been worked in a down-dip direction below the water table. Unless entry into workings is feasible, voids which occur in the worked-out seam or in the overlying collapsed ground may be detected during investigation. However, such voids might remain undetected. If workings are identified in a dipping sequence, it may be practicable to move major structures so that the foundations lie outside the limits of influence of mining or on more competent members of the cyclothem. For example, in a development on Brislington Hill, Bristol, a multi-storey block was sited over old workings varying from the surface to 15 m depth. It would have been practicable to have relocated the block onto a sandstone layer below the worked Trench Seam. However, for reasons of site planning, it was preferred to construct the block over the worked-out ground and form the foundations by deep excavation.

CONSTRUCTION OVER OLD WORKINGS SHOWN TO BE STABLE

Buildings may be constructed over old mine workings provided the foundations are designed with the ground conditions in mind or the workings are stable.

Where the allowable bearing pressure of the foundation materials has been reduced by undermining, it is possible for a raft foundation to be adopted. A raft has the advantage that it will span across weaker and more deformable zones in the foundation, and can spread the weight of the structure well outside the geometrical limits of the building. Rafts are normally expensive foundations for major structures and, for this reason, tend to be used only where no alternative solution exists. For low buildings, of about four storeys height, it is occasionally possible to use an external reinforced ring beam with a central lightly reinforced raft as a practical and economic foundation.

Before workings can be demonstrated to be stable it is normally essential to carry out some type of *in-situ* inspection of the ground conditions. The most common case is where a seam has been worked on the room and pillar pattern, and the question arises as to the adequacy of the remaining pillars in providing support for the additional surcharge to be provided by the new buildings. Surface damage can arise from a variety of forms of ground collapse associated with room and pillar workings: (a) pillar deterioration and failure, (b) bearing-capacity failure of pillar into seatearth floor and (c) void migration through rock and overburden cover. The load being carried by the pillar prior to construction can be calculated knowing the percentage extraction of the coal seam. The transfer of the weight of the building into the residual pillars is normally calculated on the assumption that the additional load acts vertically downwards and there is no lateral spreading of load. This assumption is based on the field recognition that void migration above rooms will cause dilation and joint-opening so reducing or preventing any lateral distribution of load. The strength of coal seams can be measured and should be related to observations of pillar strength *in situ*[11]. Considerable attention needs to be given to the possible effect of deterioration (either past or future) on pillar strength and the possible need for pillar protection and the construction of supplementary support. The strength and deformability of seatearths can be variable and many seatearths are subject to loss of strength when saturated. Although it is common practice to backfill workings, when open and accessible, consideration always deserves to be given to the possibility that the pillars are left as the primary means of continued support. The actual increase in pillar stress arising from construction of a building may be, proportionally, relatively small.

Considerable attention has been given in the Pittsburgh area of the USA to the possibility of leaving the coal in place as the means of support (Fig. 8.1)[12]. For example, a school was planned to be built over the Pittsburgh coal seam at an average depth of 15·5 m. After draining the mine, it was decided to relocate the school over an area of 30% coal extraction (Fig. 8.2)[12]. A further example arose where a prestressed water tank was to be built over the same seam which was at a depth of 28 m. Although the intention was for the tank to be supported on large-diameter piers, drilling of the first hole provided access and demonstrated that there had been 52% extraction. After careful evaluation of the pillar size, strength and condition, roof properties and the risk of future deterioration due to flooding and other causes, it was decided that the pillars would sustain stresses significantly higher than those associated with the proposed loading.

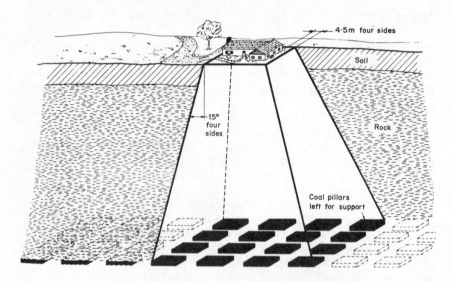

8.1. Method of supporting a structure (after Gray and Meyers[12])

In cases where the residual pillars are being relied upon to carry the structural loads, it is normal practice in the USA to assume that the area of influence is that obtained by projecting downwards from the perimeter of the area to be supported at an angle of 15° to the vertical to the level of the top of the seam.

FOUNDATION CONSTRUCTION BELOW THE LEVEL OF WORKING

If the rock above old mine workings is shown, or suspected, to be incapable of sustaining the building loads then it may be possible to support the structure on foundations carried down to below the level of the old workings. In such circumstances, it is always necessary to consider the relative economics of deep, or otherwise complex, foundations as compared with the application of ground treatment. The evaluation of these alternatives requires a consideration of the structure and its function, the overburden, the bedrock type and structure, the degree of extraction, the extent of collapse and future stability of the workings, and the ground-water conditions.

157

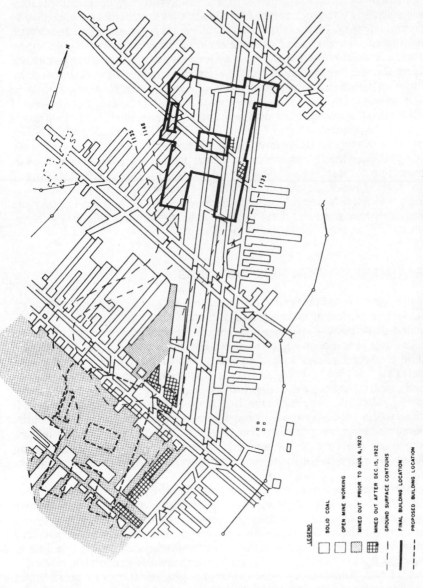

LEGEND

SOLID COAL

OPEN MINE WORKING

MINED OUT PRIOR TO AUG. 6, 1920

MINED OUT AFTER DEC. 15, 1922

GROUND SURFACE CONTOURS

FINAL BUILDING LOCATION

PROPOSED BUILDING LOCATION

8.2. Plan of mine beneath school (after Gray and Meyers[12])

In general terms, there are two main methods of foundation construction (apart from the use of a near-surface raft referred to earlier) namely, bulk excavation to the seam and the use of localised footings or sleeved piles drilled to below the level of the seam. Bulk excavation is commonly an economic solution particularly to depths of up to 7 m and on sloping sites. Such excavation may be carried out rapidly, particularly if the rocks overlying the seams are dominantly argillaceous and have been fragmented by subsidence fracturing. At progressively greater depths, the most effective alternative is the use of sleeved large-diameter piles. An important factor which may control the use of piles is the extent to which hard sandstone layers, forming an impediment to drilling, are present above the worked coal seam. The main hazard in the use of such piled foundations is that the piling process, or other changes, might encourage new collapse in the workings and consequential subsidence. Downward movement of the bedrock can result in drag-down on the piles and consequential risks of serious overloading of the pile or its foundation. Provision of a tubular steel sleeve around the pile or the bitumen coating of the pile have both been used to remove or reduce the bond between the pile and the surrounding rock. Possibly a greater hazard arises in the case of lateral transfer of load causing buckling of the piles as a result of internal toppling within the collapsing rock mass. Nevertheless, large-diameter bored piles, providing they are adequately reinforced and sleeved, can provide, in the appropriate geological circumstances, an economical foundation solution.

TREATMENT OF OLD MINE WORKINGS

Where the old workings present a potential hazard to existing or new structures and it is impossible or inappropriate to carry foundations below the workings, ground treatment is commonly instituted. Such treatment generally involves a voids-filling operation in and above the level of the workings. If the workings are still in good condition and little or no deterioration of the roof has taken place, the voids can be infilled by granular materials, such as sand or fly-ash (Fig. 8.3). In such cases, the granular materials provide lateral support and protection to the existing pillars and also support the roof, so reducing the risk of void migration. If access to the workings can be obtained from the surface, the roof is temporarily shored and walls are built to prevent excessive displacement of the bulk infilling from zones where support is required. The filling materials are commonly implaced by hydraulic stowage, providing there is reasonable opportunity for drainage.

In the case of most mine workings, partial collapse has occurred and voids may exist at both seam level and in the roof. Theoretically, void migration can extend above the workings for a distance of at least five times the excavation height. If there are a series of seams from which coal or associated minerals have been extracted, then the total height of the void migration can be considerable. In consequence, it is common practice to carry out a consolidation operation in which a mass of rock below the intended structure is grouted. It may be possible, if a specific pattern of partially collapsed workings is known to be present, to carry out a bulk voids-filling operation in addition to the consolidation grouting. If a continuous series of workings are still in existence, there is then a risk that grout will be lost outside the limits of the zone required to be treated. Artificial barriers can be created by placing pea gravel down large-diameter boreholes and then grouting the resulting gravel mound. Whether the operation is designed as a consolidation or bulk filling process, the grout mixes are typically low-cost, involving a significant proportion of *PFA*, sand or quarry scalpings. The volume over which treatment is required can be determined by using an angle of draw of 35° to the vertical between the outer limits of the structure and the base of the workings. Inevitably, such treatment of old mine workings requires careful planning and is relatively expensive; costs are typically of the order of tens of thousands of pounds for a multi-storey structure.

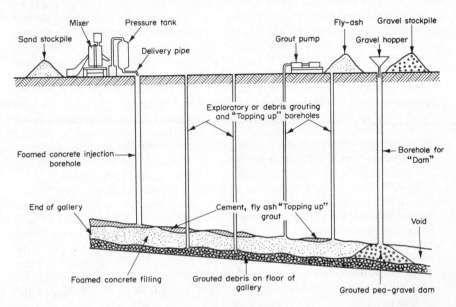

8.3. Schematic layout of plant and boreholes for filling workings beneath Sheffield College of Technology (after A. C. Scott (1957))

Bulk grouting will result in a significant change in mass permeability and consideration must be given to the implications of such treatment on ground-water flow. The possibility exists that adits, constructed to drain the workings and still effective, may be inadvertently blocked thus causing a change in water regime, possibly associated with a rising water table.

Attention should be drawn to an alternative method which has also been adopted, primarily in the USA[12]. This technique involves the drilling of 150 mm boreholes to the level of the working and then injecting grout at the level of the seam. If a mined-out void is encountered, then gravel is added until a mound is formed which supports the roof; the gravel is then grouted. Otherwise, grouting is carried out from the borehole which is left full of grout at the end of grouting. The basic concept is to create a grouted column composed of the cemented borehole surrounded by a zone of grouted rock; the gravel mound would span across voids. By this means a crude form of pile is created *in situ*. The technique is possibly most successfully applied where roof subsidence has not occurred to any major extent. The published costs indicate that there is little economic difference between this process and consolidation grouting based on closely spaced boreholes.

Special Problems

HILLSIDE STABILITY

Excavation into hillsides underlain by dipping sequences may result in displacement unless adequate support is provided at an early stage. At the site of the Hallamshire hospital, Sheffield, a face 13 m in height was cut into a sequence of mudstones with thin sandstones and seatearths dipping at 18–19° into the excavation. The high part of the rock face was stabilised by the installation of several rows of cables placed in boreholes at 4 m centres and stressed to 2·6 MN against vertical concrete pillars. Subsidiary support was provided by horizontal walings formed by lengths of sheet pile and intermediate widely spaced vertical planks. During the excavation of another part of the site a bedding plane slide occurred; large shear box tests revealed that ϕ'_p and ϕ'_r for the natural bedding surfaces were 17° and 16°, respectively. Rock sequences which contain such argillaceous, poorly cemented rocks will be subject to bedding-plane slip during folding. It has been recognised by Stimpson and Walton[13] that sheared clay 'mylonite' bands in the Coal Measures may have angles of shearing resistance as low as $\phi'_p = 11°$ and $\phi'_r = 10°$. In such circumstances, deep excavations into dipping Coal Measures, below the water table, should be provided with artificial support, even at relatively low angles of dip. By way of example, in connection with the construction of the Eldon telephone exchange, Sheffield, an excavation ranging in depth from 9 to 6 m was carried out into a hillside. The site was typically underlain by 1·5 m of fill, up to 6 m of mudstone and siltstone, nearly 5 m representing a former zone of coal working and a light grey siltstone of undetermined thickness; the rocks were folded into a gentle syncline. During the progressive excavation, the sides of the cut were supported by wider beams and 0·5 MN cables inclined at 30° to the horizontal and anchored into the thick siliceous siltstone. The average rate of pumping from the site, which at its deepest was 3 m below the water table, was in excess of 10^6 litre/day.

INFLUENCE OF DEEP MINE WORKINGS

The problems which arise specifically from shallow mine workings, considered previously, have an important influence on foundation conditions, but it is, however, common practice to ignore the possible effects of former deep mining. It is generally recognised that subsidence arising from deep mining is normally completed within 10–30 years. After such a period, any hazards to foundation construction are believed to be eliminated by cessation of settlement. However, during undermining, the rock mass undergoes a volume increase because in fully extracted areas the subsidence is less than the seam thickness. This dilation results from fissure and fracture opening and a consequential increase in secondary porosity and permeability[14]. The most marked influence of undermining occurs on the overlying rocks immediately above the former limits of extraction. The residual strains remaining in the rock after mining ceases can result in the creation of fractures, some of which may extend to the ground surface as large tension cracks. In consequence, undermining must result in an inevitable deterioration in the foundation properties of rocks thus influenced. The extent of this deterioration is, however, difficult to predict and must represent a combination of factors including seam thickness, depth of the seam, lithology of the overlying rocks and mining practice. If the rock exposed at rockhead is a sandstone, it is likely that the fractures will be relatively wide and set far apart. Such fractures can be treated, when exposed, by excavation below the broad foundation level and backfilling with concrete. In the case of mudstones and shales, the fissuring is likely to be more penetrative, the rock will be less compact and there will be a consequential reduction in the allowable bearing pressure for the rock mass. The greatest uncertainty in assessing foundation conditions arises when the bedrock is hidden below thick overburden, and the foundation is to be of piled construction.

In areas within which deep mining has occurred, and good records exist, it is possible to reconstruct the residual pattern of strains. Such techniques have been used, for example, in the vicinity of the tips of Aberfan in south Wales and the Tyne tunnel. Having established this strain pattern, it is possible to appraise the foundation conditions in the light of knowledge of the location, distribution and magnitude of the past strains[15].

OLD MINE SHAFTS

Major problems can arise when there is an old shaft within the vicinity of a proposed development. In many cases the exact location of the shaft is not known and the precise position needs to be determined by investigation. Such exploration and subsequent treatment of the shaft must be carried out with the knowledge and agreement of the National Coal Board, who will normally provide a detailed specification for the procedure to be adopted. The main hazard associated with a shaft is that material will fall into it resulting in rapid subsidence and loss of support to the ground surface or adjacent foundations. It may be practicable to relocate the various structures around the site of a known shaft so that no potential hazard exists. However, in the case of a deep shaft or one that is known to be still open, or where the precise location is uncertain, then the appropriate exploration and treatment must be carried out. If the top of the shaft is not visible, it should be exposed by a drag-line securely anchored outside the limits of the expected shaft position. However, if this technique is impracticable, the shaft is normally explored by a borehole drilled down the shaft centre to bedrock; in addition, a further borehole is located next to the shaft to determine the elevation of rock head. The treatment of a disused shaft may be varied depending upon:

1. The thickness and properties of the overburden.
2. The presence and type of filling which may already be in the shaft.
3. Whether or not noxious gases are present.

If the cover of overburden is not excessive and the shaft is open, it can be filled by suitable granular material. However, if as is more likely, the shaft is partially filled by debris, possibly lodged on staging, then the voids must be proved by drilling. The voids are then progressively backfilled by pea gravel, where practicable, starting at the base of the shaft. In both of these cases the shaft is then covered by a square reinforced-concrete raft, not less than 0·6 m thick and with an edge length not less than twice the shaft diameter, located at rockhead.

On the other hand, if the depth of overburden is excessive, the shaft should be backfilled with gravel to ground level and then grouted with a cement-fly ash grout.

The top of the shaft should be covered by either a raft below ground level or a concrete plug in the shape of an inverted truncated cone. The back filling of an old shaft can be a difficult task, particularly if the shaft is partially filled. For example, in the case of a shaft at Kingswinford, Staffordshire, repeated collapse of wooden stagings occurred during the progressive drilling-in of casing which was to be used in the filling process.

One of the most straightforward methods of coping with mine shafts is simply to avoid locating structures in their immediate proximity. For example, a group of multi-storey blocks was moved some 20 m away from three shafts at Torre Road, Leeds, but nevertheless the shafts were capped as a precaution. The minimum safe distance for siting buildings from open or poorly filled shafts is determined by the condition and thickness of the overburden. It is recommended that this minimum distance should be twice the overburden thickness up to a depth of 15 m, unless the overburden is exceptionally weak. Protective sheet pile or concrete walls can be constructed around shafts. In the case of a development at West Bromwich, the bedrock was composed of the Keele Beds which were overlain by 25 m of glacial deposits and fill. Two shafts are present on the site, the Heath Pits shaft with a depth greater than 280 m and the Engine Pit shaft with a depth of about 190 m. The original recommendations for coping with the shafts included filling them if they were still open, or else infilling them with artificial replacement where required; all the shafts had to be capped. In view of the depth of the shafts there was a risk that consolidation of the filling material placed in them would take an excessive time. Because of the site difficulties and the presence of ground-water, it was decided to treat the Heath Pits shaft by a combination of filling it and constructing an external reinforced-concrete diaphragm wall to safeguard the possible failure of the vulnerable part of the shaft in the glacial overburden. The shaft was found to be filled a few metres below rock head. The lower part was then backfilled by *PFA*–cement grout. This was then covered by a reinforced-concrete cap at some 15 m below the surface. The remainder of the shaft was filled with granular material and covered by a concrete shaft capping. The external protection took the form of a 600 mm wide diaphragm wall, octagonal in plan, excavated a minimum of 3 m into rock. The Engine Pit shaft, in contrast, was treated by *PFA*–cement grout over its whole depth.

INFLUENCE OF PYRITES

Many rocks of the Coal Measures, particularly shales, mudstones and coals, contain iron sulphides, in the form of pyrite or marcasite. Weathering of such rock types results in the formation of ferrous sulphate and sulphuric acid; the sulphate may further combine with water to yield limonite and more sulphuric acid. If calcium carbonate is present, as in thin limestones or marine shales in particular, gypsum will be formed thus leading to a volume increase, disruption of the rock and consequential weakening. The essential reactions are as follows:

$$2FeS_2 + 2H_2O + 7O_2 \rightarrow 2FeSO_4 + 2H_2SO_4$$

$$FeSO_4 + 2H_2O \rightarrow Fe(OH)_2 + H_2SO_4$$

$$H_2SO_4 + CaCO_3 + 2H_2O \rightarrow CaSO_4 \cdot 2H_2O + H_2CO_3$$

The sulphate ions are aggressive to concrete, reacting with the tricalcium aluminate in cement to form soluble calcium sulpho-aluminate. The most severe effects of sulphide decay occur in situations such as shale tips or rocks influenced by mining subsidence, where there is a high permeability and adequate through-flow of oxygenated water. Temperature and bacteria may influence the process. In such circumstances several per cent of sulphate has been recorded in shales and concentrations of several thousand parts per million in pore water. Apart from the deleterious influence on concrete, sulphides can give rise to structural damage by virtue of expansive action during decay. This process has not been recorded to any great extent in the UK but is well known in the coalfields of the Appalachians. The existence of this hazard arises largely from the use of unreinforced floor slabs for low structures, thus resulting in heaving and consequential damage. The formation of gypsum (and melanterite, an insoluble complex sulphate) results in an increase in volume of eight times the original sulphide and consequential stresses of possibly up to 0.5 MN/m². Experience in the USA has suggested that a sulphide content of 0.1% is adequate to cause deleterious expansion.

The occurrence of pyrites in bedrock or fill materials is normally regarded as hazardous in foundation construction in the UK. The lack of occurrences of the same problems in the UK that occur in identical rocks in the USA is probably the result of established foundation practice involving either excavating or avoiding such suspect materials. Alternatively, the expansion may still occur either into loosely backfilled foundation excavations and sub-basement openings, or in such a way that its influence is restrained by high foundation loadings.

REFERENCES

1. Price, D. G., Malkin, A. B. and Knill, J. L., 'Foundations of Multi-Storey Blocks with Special Reference to Old Mine Workings', *Quart. J. Eng. Geol.,* 1 No. 4, 271–322 (1969)
2. Franklin, J. A., Broch, E. and Walton, G., 'Logging the Mechanical Character of Rock', *Trans. Inst. Min. Met.,* **81**, A1–9 (1971)
3. *Foundations Code of Practice,* CP 2004, B.S.I., 2nd edn (1972)
4. Spears, D. A. and Taylor, R. K., 'The Influence of Weathering on the Composition and Engineering Properties of *in-situ* Coal Measure Rocks', *Int. J. Rock Mec. Min. Sci.,* 9 No. 6, 729–756 (1972)
5. Meigh, A. C., 'Foundation Characteristics of the Upper Carboniferous Rocks', *Quart. J. Eng. Geol.,* 1 No. 2, 86–113 (1968)
6. Smith, D. W., 'New Scotswood Bridge', *Proc. Inst. Civ. Eng.,* **42**, 217–250 (1969)
7. Thorburn, S., 'Large Diameter Piles Founded on Bedrock', in *Symposium on Bored Piles,* Institution of Civil Engineers, London (1966)
8. Crutchlow, S. B., 'The Foundation Properties of Upper Carboniferous Shales', *Proc. 1st Int. Conf. Rock Mechs.,* 1, Lisbon (1966)
9. Taylor, R. K., 'Site Investigations in Coalfields—the Problem of Shallow Mine Workings', *Quart. J. Eng. Geol.,* 1 No. 2, 115–133 (1968)
10. Price, D. G., 'Engineering Geology in the Urban Environment', *Quart. J. Eng. Geol.,* 4 No. 3, 191–208 (1971)
11. Bienawski, Z. T., '*In situ* Strength and Deformation Characteristics of Coal', *Eng. Geol.,* 2 No. 5, 325–340 (1968)
12. Gray, R. E. and Meyers, J. F., 'Mine Subsidence and Support Methods in the Pittsburgh Area', *J. Soil Mech. and Found. Div.,* ASCE, **96** No. SM4, 1267–1287 (1970)

13. Stimpson, B. and Walton, G., 'Clay Mylonites in English Coal Measures. Their Significance in Open-cast Slope Stability', *Proc. 1st Int. Cong. Int. Assoc. Eng. Geol.*, **2**, Paris (1970)
14. Knill, J. L., 'The Engineering Geology of Old Mine Workings', Paper No. 39, The Midlands Soil Mechanics and Foundation Engineering Society (1972)
15. Knill, J. L., 'Rock Conditions in the Tyne Tunnel, North Eastern England', *Bul. Assoc. Eng. Geol.*, **10** (1973)

Index